Fascinating Life Sciences

This interdisciplinary series brings together the most essential and captivating topics in the life sciences. They range from the plant sciences to zoology, from the microbiome to macrobiome, and from basic biology to biotechnology. The series not only highlights fascinating research; it also discusses major challenges associated with the life sciences and related disciplines and outlines future research directions. Individual volumes provide in-depth information, are richly illustrated with photographs, illustrations, and maps, and feature suggestions for further reading or glossaries where appropriate.

Interested researchers in all areas of the life sciences, as well as biology enthusiasts, will find the series' interdisciplinary focus and highly readable volumes especially appealing.

Lauren Moretto • Joanna L. Coleman
Christina M. Davy • M. Brock Fenton
Carmi Korine • Krista J. Patriquin
Editors

Urban Bats

Biology, Ecology, and Human Dimensions

 Springer

Editors
Lauren Moretto
Vaughan, ON, Canada

Christina M. Davy
Department of Biology
Carleton University
Ottawa, ON, Canada

Carmi Korine
Mitrani Department of Desert Ecology
Ben-Gurion University of the Negev
Midreshet Ben- Gurion, Israel

Joanna L. Coleman
Department of Biology
Queens College at the City University of
New York
Flushing, NY, USA

M. Brock Fenton
Department of Biology
University of Western Ontario
London, ON, Canada

Krista J. Patriquin
Department of Biology
Saint Mary's University
Halifax, NS, Canada

ISSN 2509-6745　　　　　　　　ISSN 2509-6753　(electronic)
Fascinating Life Sciences
ISBN 978-3-031-13175-2　　　　ISBN 978-3-031-13173-8　(eBook)
https://doi.org/10.1007/978-3-031-13173-8

Cover illustration: Photograph designed by Nathan W. Fuller

This Springer imprint is published by the registered company Springer Nature Switzerland AG
The registered company address is: Gewerbestrasse 11, 6330 Cham, Switzerland

Preface

The current geological epoch, the Anthropocene, is characterised by human activities that are so extensive and intense that they are leaving durable signals in stratigraphic material. At the same time, wildlife must now contend with new challenges in the Anthropocene, and previous work has shown that bats are no exception (e.g., Voigt & Kingston's 2016 *Bats in the Anthropocene*). Nowhere is this truer than in cities, which are the focus of ***Urban Bats: Biology, Ecology, and Human Dimensions***. Bats in urban areas experience the pressures of profound and ongoing land cover change and other diverse challenges. Research on urban bats in recent decades indicates that most bat species are negatively impacted by urbanisation. Yet some species of bats succeed and even thrive in cities and towns. These observations have inspired questions about bats in relation to urbanisation. Which traits and behaviours equip bats for urban success? What features of urban areas increase the likelihood that bats will successfully persist there or even colonise new areas? And how does the success of urban bats affect human urbanites? To address these questions, ***Urban Bats: Biology, Ecology, and Human Dimensions*** builds on the foundation of knowledge of bats in cities and adjacent areas, and aims to inspire future research and promote the conservation of urban bats.

Intended primarily for environmental managers and an academic audience, this book brings together authors and researchers from around the world to explore the interactions between bats and urban environments. Thirteen chapters, organised into three parts, explore these interactions through case studies or a summary of the existing literature.

Part I, What is an Urban Bat? Morphological, Physiological, Behavioural, and Genetic Adaptations, reviews the biology of urban bats. Chapter 1 addresses traits related to stress and thermal physiology that might confer an advantage to bats using urban areas. Chapter 2 explores the genetic impacts of urbanisation on the family Pteropodidae, and Chap. 3 follows with a look at possible pre-adaptations of molossid bats to life in cities. Phenotypic changes in urban bats are also explored through their interactions with their ecto- and endoparasites (Chap. 4). Finally, Chap. 5 examines the reproductive success and social structure of urban, frugivorous bats.

Part II, How do Bats Inhabit Urban Environments? Uses of Artificial Roosts, Aerial Habitats, and Green Spaces, examines the interactions between urban bats and their environments. Chapters 6 and 7 explore bats' use of natural and artificial habitats in cities, with foci on bat box structures, urban green spaces, and water bodies (blue spaces). Chapter 8 discusses some of the particular challenges that bats face in navigating urban, aerial habitats. Chapter 9 investigates how insectivorous bats in the Brazilian Atlantic Forest area are responding to recent and rapid growth of nearby cities.

Part III, How do Bats and Humans Interact in Urban Environments? Human Perceptions, Public Health, and Ecosystem Services of Bats, offers insight into relationships between urban-dwelling bats and humans, including human knowledge and attitudes (Chap. 10), public health considerations (Chap. 11), and bat-mediated ecosystem services (Chap. 12).

Finally, Chap. 13 offers a "big picture" perspective on urban bats, reflecting on the information presented in previous chapters. In this chapter, we conclude the book by considering how urban bats' responses to urbanisation compare to those of other, urban-dwelling mammals. Finally, we identify key knowledge gaps that can be prioritised for future research and conservation efforts.

Vaughan, ON, Canada   Lauren Moretto
Flushing, NY, USA   Joanna L. Coleman
Ottawa, ON, Canada   Christina M. Davy
London, ON, Canada   M. Brock Fenton
Midreshet Ben- Gurion, Israel   Carmi Korine
Halifax, NS, Canada   Krista J. Patriquin

Acknowledgements

We are grateful to Burton K. Lim for guidance in the early stages of this project, and to all who provided support during the writing and editing process.

Contents

Part I   What is an Urban Bat? Morphological, Physiological, Behavioural, and Genetic Adaptations

1   **Stress Physiology, Foraging, and Ecophysiology of Bats in Urban Environments**.   3
Carmi Korine, Phillip J. Oelbaum, and Agustí Muñoz-Garcia

2   **Genetic Impoverishment in the Anthropocene: A Tale from Bats** . . .   19
Balaji Chattopadhyay, Kritika M. Garg, Rajasri Ray,
Ian H. Mendenhall, and Frank E. Rheindt

3   **Are Molossid Bats Behaviourally Preadapted to Urban Environments? Insights from Foraging, Echolocation, Social, and Roosting Behaviour**.   33
Rafael Avila-Flores, Rafael León-Madrazo, Lucio Perez-Perez,
Aberlay Aguilar-Rodríguez, Yaksi Yameli Campuzano-Romero,
and Alba Zulema Rodas-Martínez

4   **Urban Bats and their Parasites**   43
Elizabeth M. Warburton, Erin Swerdfeger, and Joanna L. Coleman

5   **Bat Societies across Habitat Types: Insights from a Commonly Occurring Fruit Bat** *Cynopterus sphinx*   61
Kritika M. Garg, Balaji Chattopadhyay, D. Paramanatha Swami Doss,
A. K. Vinoth Kumar, and Sripathi Kandula

Part II   How do Bats Inhabit Urban Environments? Uses of Artificial Roosts, Aerial Habitats, and Green Spaces

6   **Bat Boxes as Roosting Habitat in Urban Centres: 'Thinking Outside the Box'**   75
Cori L. Lausen, Pia Lentini, Susan Dulc, Leah Rensel,
Caragh G. Threlfall, Emily de Freitas, and Mandy Kellner

7   Aerial Habitats for Urban Bats 95
       Lauren A. Hooton, Lauren Moretto, and Christina M. Davy

**8   City Trees, Parks, and Ponds: Green and Blue Spaces as Life
       Supports to Urban Bats** 107
       Lauren Moretto, Leonardo Ancillotto, Han Li, Caragh G. Threlfall,
       Kirsten Jung, and Rafael Avila-Flores

**9   Assessing the Effects of Urbanisation on Bats in Recife Area,
       Atlantic Forest of Brazil** 123
       Enrico Bernard, Laura Thomázia de Lucena Damasceno,
       Alini Vasconcelos Cavalcanti de Frias, and Frederico Hintze

**Part III   How do Bats and Humans Interact in Urban Environments?
             Human Perceptions, Public Health, and Ecosystem Services
             of Bats**

10   Human Dimensions of Bats in the City 139
       Leonardo Ancillotto, Joanna L. Coleman, Anna Maria Gibellini,
       and Danilo Russo

11   Urban Bats, Public Health, and Human-Wildlife Conflict 153
       Christina M. Davy, Arinjay Banerjee, Carmi Korine, Cylita Guy,
       and Samira Mubareka

12   Ecosystem Services by Bats in Urban Areas. 167
       Danilo Russo, Joanna L. Coleman, Leonardo Ancillotto,
       and Carmi Korine

**13   The Big Picture and Future Directions for Urban Bat
       Conservation and Research** 181
       Krista J. Patriquin, Lauren Moretto, and M. Brock Fenton

Index 189

Part I
What is an Urban Bat? Morphological, Physiological, Behavioural, and Genetic Adaptations

Chapter 1
Stress Physiology, Foraging, and Ecophysiology of Bats in Urban Environments

Carmi Korine, Phillip J. Oelbaum, and Agustí Muñoz-Garcia

Abstract Urban environments alter various physiological responses of sympatric animals. In this chapter, we provide an overview of stress physiology and the responses of bats to different urban stressors, such as light and noise pollution. We suggest future directions of research connecting urbanisation and stress responses of bats, whose life history traits determine an idiosyncratic response to urbanisation. We review how foraging behaviour and the physiological ecology of bats vary with the urban environment and present data on the effect of roost microclimate on metabolism and water balance of bats. We discuss these findings under an evolutionary lens and conclude that synanthropic species of bats possess preadaptations. These adaptations include resilience to urban stressors, fast flight, use of sheltered roosts, and relatively low metabolic rates to survive and thrive in urban habitats.

Keywords Stress physiology · Noise and light pollution · Roost energetics · Water balance

Supplementary Information The online version contains supplementary material available at [https://doi.org/10.1007/978-3-031-13173-8_1].

C. Korine (✉)
Mitrani Department of Desert Ecology, Ben-Gurion University of the Negev, Midreshet Ben- Gurion, Israel
e-mail: ckorine@bgu.ac.il

P. J. Oelbaum
Department of Cell and Systems Biology, University of Toronto, Toronto, ON, Canada

A. Muñoz-Garcia
Department of Evolution, Ecology and Organismal Biology, Ohio State University at Mansfield, Mansfield, OH, USA

1 Introduction

In the last two centuries, the human population has increased exponentially, with a massive migration of people from rural to urban centres. Currently, 25% of the world's population lives in cities with over one million inhabitants.[1] Modern cities are a relatively new environment for animals, and many species have found opportunities to exploit new niches or have traits that make them well suited to coexist with humans. These species can be described as synanthropic. Among 1450 known species of bats, few can be considered synanthropic, and most bat species are negatively affected by urbanisation [1, 2]. Environmental conditions in urban habitats, such as light and noise pollution, possibly air and water pollution, and low diversity of suitable roosts, may be detrimental to bats. Other factors, including increased ambient temperature (T_a), high availability of food through the introduction of ornamental and invasive species of plants and animals, high density of food patches, abundance of drinking water sources, and high availability of certain roost types, may favour the presence of bats in urban environments [3].

In this chapter, we present studies that test the stress responses of birds and mammals to urban disturbances and review how alteration of environmental variables in urban environments may affect the stress response, foraging behaviour, and physiological ecology of urban bats. Further, we present data on the effect of roost microclimate on metabolism and water balance of bats and synthesise these physiological and dietary responses. We discuss how the physiology of bats might be affected by the urban conditions and conclude that synanthropic species of bats may possess preadaptations to survive and thrive in urban habitats.

2 Stress Physiology

The stress response has been defined as an organism's behavioural or physiological reactions to a variety of external stressors (e.g. pesticide levels, handling, captivity, predation, heat load) and internal factors such as parasite load or starvation [4, 5]. Stressors may have short- or long-term effects, and animals may experience both types of effects during their lifetime. Responses to stressors are regulated by the hypothalamic-pituitary-adrenal (HPA) axis, through the elevation of glucocorticoid (GC) hormone levels. Other neuroendocrine systems, such as the secretion of melatonin by the pineal gland, also interact with the HPA axis. These hormones are responsible for maintaining the energetic balance needed to cope with stressors, and under normal conditions, they regulate growth, circadian activity, reproduction, and immune function in animals [6]. However, high levels of stress hormones are known to inhibit the reproductive system and depress immune responses [6].

---

[1] https://ourworldindata.org/urbanization, https://data.worldbank.org/indicator/EN.URB.
MCTY.TL.ZS)

In bats, studies have shown that cortisol is the primary GC occurring in plasma. However, available data are scarce because only 15 species have been studied (Table S1.1). Other physiological responses may be associated, either directly or indirectly, with stress hormones, including changes in haematological variables, depression of immune function, increases in oxidative stress, and decreases in body condition [6].

Stress responses are commonly estimated by measuring the concentrations of baseline plasma GC; however, this method is invasive, not suitable for use in small animals, and only reflects variation in GC concentrations over the preceding hour. For that reason, those who research wildlife stress physiology often use measurements of faecal glucocorticoid metabolites (FGM) and cortisol levels in urine or hair [Table S1.1 for studies on bats; 6–10]. Faecal or urine concentrations of cortisol reflect average daily levels of the hormones, whereas cortisol concentrations in hair reflect circulating levels over periods of several weeks or months [5]. The FGM method is recommended when studying insectivorous bats because they are small and sensitive to handling. However, the FGM method has certain limitations, such as the inability to assign collected samples from daily roosts to specific individuals or sexes. When studying fruit bats, plasma GC levels may also be obtained (Table S1.1), as long as the whole blood sampling procedure takes less than 3 min of handling [11]. Furthermore, because GC concentrations follow daily rhythms [11], it is essential to measure baseline concentrations of the hormone before inferring the effects of stressors.

Stress responses of animals are related to many external factors, such as seasonality, city size, and experimental set-up [6]. These factors reflect the stressor type and magnitude, and the duration of exposure to the stressor. In addition, stress responses are modulated by intrinsic factors, such as sex, age, reproductive status, food availability and quality, degree of plasticity to the response, and genetic background [12, 13]. The complex relationships between stress hormones and extrinsic and intrinsic factors may explain the contradictory physiological responses of wildlife to urbanisation. Indeed, reviews on birds [13] and several other vertebrate taxa [6] have found that baseline or stress-induced plasma GC levels show no consistent pattern in the short-term responses to chronic stress. However, these studies also reported that, across all vertebrate taxa analysed, anthropogenic disturbances are significantly associated with increased FGM concentrations. In a recent meta-analysis across a range of vertebrate taxa, Iglesias-Carrasco et al. [14] found no significant effect of urbanisation on corticosterone or cortisol concentrations in plasma, faeces, hair, or feathers. The lack of associations between stress hormone levels and the magnitude of environmental stressors could reflect differences in assay methodologies [14], specific measures of physiological responses (i.e. GC levels in hair vs. blood), types of stressors to which the animals were exposed, and intrinsic factors that affect the response to the stressor (e.g. age, reproductive status). These contradicting physiological responses should therefore be considered when evaluating the stress responses of bats to urbanisation.

Studies of animals in natural habitats are mainly limited to short-term stress responses, while life in urban environments also entails exposure to chronic stressors, such as anthropogenic noise, artificial light at night (ALAN), and air and water pollution. Stressors imposed on wildlife in urban habitats force animals to modify their behavioural and physiological responses in a way that either avoids or allevi-ates the stress [15, 16]. In the next section, we summarise known physiological stress responses to specific urban stressors that may affect bats.

3   Extrinsic Factors that Affect the Stress Response in Bats

Bats are highly mobile, mostly social, small, and sensitive to handling. As such, their physiological responses to stress may be particularly complicated to study – perhaps this explains why only a few stress physiology studies have focused on bats (Table S1.1). Most such studies have aimed either to evaluate basic concentrations of stress hormones in blood or faecal samples or to test how intrinsic factors, such as sex, restraint, seasonality, and reproductive status, may affect the stress responses (Table S1.1). Only two studies have addressed the stress response of bats in direct relation to urbanisation [15, 17], while a third examined the effect of contaminated water [18], and a fourth considered the effect of wind turbines [19].

3.1   Stress and Light Pollution

Artificial light at night (ALAN) is among the most striking environmental changes associated with the expansion of urbanisation [20]. Artificial light at night disrupts the natural spatial and temporal patterns that regulate light-dependent biological processes in many organisms, including bats. In most bat species, ALAN elicits avoidance behaviour [21–23] except for a few opportunistic species that forage on insects or fruits [21]. Exposure to ALAN can eliminate the natural circadian rhythm of GC release and increase GC concentrations in captive [24] and free-living ani-mals [25] across different taxa, but studies on urban bats are yet to be performed. Nevertheless, Injaian et al. [26] did not find a general relationship between ALAN and baseline stress-induced corticosterone in birds and reptiles in their review. In mammals, ALAN decreased GC concentrations, immune function, and body mass [26–27]. However, these studies were mainly done under laboratory conditions and high light intensities. No evidence of a decrease in FGM levels, body mass, nor survival probability were found when conducted with free-living bank voles (*Myodes glareolus*) [28].

3.2   Stress and Noise Pollution

There is overwhelming evidence that anthropogenic noise constitutes an ecological pollutant, with many deleterious effects on wildlife [29], although known physiolog-ical responses of wildlife are variable. Noise from motorways seems to increase

FCM concentrations of free-living wood mice [*Apodemus sylvaticus*; 30]. Brearley et al. [31] reported that cortisol concentrations in the hair of squirrel gliders (*Petaurus norfolcensis*) were elevated in animals that lived in edge habitats along major roads that were noisy and with limited availability of hollow nests. In contrast, Łopucki et al. [32] reported that corticosterone concentrations in faecal samples from a rural population of striped field mice (*Apodemus agrarius*) were not different from those of an urban population exposed to various anthropogenic stressors, including noise; their findings might reflect a habituation response to urban noise.

For most bat species, echolocation is a crucial activity for navigating and detecting resources, such as water and food. Thus, noise pollution may alter echo reception and processing in bats, impairing the basic functioning of their echolocation abilities, and may interfere with drinking behaviour [33] and passive detection of prey-generated acoustic cues [34–35]. Allen et al. [15] found that nonreproductive or lactating female Mexican free-tailed bats (*Tadarida brasiliensis*) roosting under bridges in Texas, United States, had lower baseline plasma GC concentrations than females roosting in caves, suggesting that bats may not be affected by noise and light disturbances. However, during pregnancy, the highest baseline GC concentrations were found in bats roosting under a bridge in a suburban area. Parry-Jones et al. [17] found that urban grey-headed flying foxes (*Pteropus poliocephalus*) in Sydney, Australia, had higher FGM concentrations than rural bats, but linked their findings to food shortages that the bats experienced during the study, not to the chronic stress of urbanisation. Wada et al. [18] and Medina-Cruz et al. [19] found no significant differences in GC concentrations between bats captured in control and disturbed sites in South River, Virginia, United States, and Oaxaca, Mexico, respectively.

4   Intrinsic Factors That Affect the Stress Response in Bats

In addition to the urban disturbances mentioned above, specific traits of bats should be considered when studying their physiological response to stress in urban environments. Most bat species are sensitive to anthropogenic disturbances [1]; however, stress responses of populations of conspecifics might vary with the degree of urbanisation. It is reasonable to predict that, within a species, stress responses will be stronger in populations that avoid urban habitats than in populations that occupy them. Additionally, a city's size, age, and composition (e.g. presence of large parks, high-rise buildings, human population density) may play a role in how bats respond to and adapt to urban stressors. To understand the magnitude of the stress responses to urbanisation, we suggest studying the stress physiology of several colonies of a sympatric species that vary in their association with urban environments or the size of cities they inhabit, considering their colony size and degree of mobility, as well as other intrinsic factors, such as diet and trophic guild (see below). In addition, incorporating the genomic variation associated with responses to increased urbanisation will facilitate testing whether stress responses result from phenotypic plasticity or represent adaptive responses to urbanisation [12].

4.1   Mobility

Bats can compensate for the short-term effects of stressors through flight. Generally, bats may be divided into fast flyers with low manoeuvrability and slow flyers with high manoeuvrability. In temperate areas, the most common urban-dwelling bats are insectivores with high-aspect ratio wings and high wing loading, which allow them to forage in open spaces, fly fast, and cover long distances [36]. These characteristics allow bats to move at high speeds with relatively low-energy expenditure, but only at the expense of reducing manoeuvrability in enclosed spaces, such as dense forests [37, 38]. These traits are positively associated with body size, which allows urban bats to feed on large prey (e.g. moths), with high-energy returns that diminish foraging time [39]. It would be interesting to test whether sympatric and more mobile bat species are less affected than sedentary bat species by short-term urban stressors and whether mobility is a (pre)adaptive trait that allows them to cope with these stressors.

4.2   Sociality

Most species of bats live in groups and are philopatric. As a result, size, composition, structure, and formation of groups may affect behavioural and physiological responses and reproductive success. Urban bats are more likely to live in larger groups compared to non-urban conspecifics because cities offer high availability of potential roosts and food sources. As such, they may experience higher levels of chronic stressors. Interestingly, Allen et al. [15] found that even though numbers of *T. brasiliensis* varied from several thousand roosting under a bridge in a suburban residential area to more than one million in one cave, the plasma GC concentrations of bats did not differ between roost sites. In the meridional serotine (*Eptesicus isabellinus*), FCM concentrations did not vary among colonies of different sizes [40]. Reeder et al. [41] suggest that being in a breeding group is chronically stressful for male pteropodids. Indeed, male large (*Pteropus vampyrus*) and little golden-mantled flying foxes (*Pteropus pumilus*) living in secondary forests in Malaysia exhibit increases in total GC concentrations during the breeding season [41]. Furthermore, GC concentrations were higher in captive males that were in the process of group formation than in males that were already in stable groups [41]. Thus, the effects of colony size on stress responses, in rural and particularly urban habitats, might be confounded with the impacts of roost quality, social challenges, and breeding. More studies that control for at least one of those factors are required to better understand these relationships.

4.3   Diet and Foraging Behaviour

Diet and foraging behaviour of urban bat populations may differ significantly from populations in their natural habitats. Changes in foraging behaviour can negatively impact bats through nutrient stress and tissue impairment and contribute to higher levels of GC plasma concentrations, causing long-term deleterious health effects [42]. Geggie and Fenton [43] found that big brown bats (*Eptesicus fuscus*) spent more time foraging in urban areas than in rural areas of Canada's National Capital Region, suggesting lower prey density in cities. In contrast, several other studies have noted decreased foraging periods in urban environments leading to the hypothesis that this may be the result of greater prey availability in cities [2]. These findings may be attributed to differences in patchiness of resources and resource accessibility [2, 36, 44]. In Mexico City, Mexico, for example, large, vegetated areas are important foraging sites for bats, especially insectivores that are drawn in by swarms of insects around streetlights [44]. Additionally, elsewhere in the Neotropics, frugivores and nectarivores exploit ornamental plants in parks and green spaces as important dietary resources [45]. We therefore suggest evaluating whether feeding in dense food patches may buffer bats against energetic stress, enabling them to cope with consistent high exposure to anthropogenic stressors. We suggest that this could be done by comparing the FGM concentrations and movements of bats between food-rich and nearby food-poor patches.

Some frugivorous and insectivorous bat species seem to preferentially seek out, or at least tolerate, urban areas. Aerial insectivores in general, and especially members of the family Molossidae, are well adapted to open spaces and more abundant in many temperate and Neotropical cities than members of other trophic guilds [44]. Urban populations of the greater noctule bat (*Nyctalus lasiopterus*) in Seville, Spain, avoided wooded areas and had smaller, but more variable, home ranges compared to populations found in a forest preserve – this shift between foraging strategies in forests and cities could affect and help to modulate stress responses in urban bats [46]. This means that if bats spend more time searching for food in forests than in cities where patches of resources are more predictable and concentrated [46], then this reduced hunting effort may modulate the impacts of other negative anthropogenic stressors.

Bats with generalist diets and wide trophic niche breadth may also be better adapted to urban life, because they can take advantage of novel resources [2]. Frugivorous bats in both the Neotropics and Paleotropics are well adapted to feed from abundant fruit sources which are popular ornamental or agricultural plants in urban areas [4, 44]. Studies that link dietary preference, diet quality, and foraging time with stress hormone levels have shown that diet quality and choice are associated with body condition and elevated stress responses in mammals [47] and birds [48]. Experiments on frugivorous bats may determine how fruit quality and dietary diversity affect FGM, considering other extrinsic and intrinsic factors. For example, the neutrophil to lymphocyte ratio, a putative stress indicator, in Neotropical frugivores increased under conditions of forest loss – a finding indicative of chronic stress [49]. These types of experiments may also show whether urban bats with generalist diets have moderated stress responses, which could indicate their degree of plasticity or preadaptation to urban habitats.

5   Ecophysiology of Bats in Urban Environments

The remarkable taxonomical and ecological diversity of bats may relate to their ability to fly. Powered flight is associated with the presence of large, vascularised wing membranes that increase the surface area-to-volume ratio, which, in turn, is associated with high rates of metabolism and evaporative water loss. Therefore, bats face potential challenges when it comes to maintaining their energy and water balances and demonstrate physiological and behavioural strategies to cope with these challenges. For example, many bats can enter daily torpor, which significantly decreases daily energy expenditure and water loss [50]. Additionally, many species of bats use roosts whose microclimates help individuals maintain energy and water balance [51]. Bats in urban habitats may therefore benefit from warmer microclimates (see below) in daily or temporary roosts and artificial water sources.

5.1   Roosts and Energetics

Bats spend most of their lives in day roosts [51]. The appropriate roost microenvironment for temperature and humidity can reduce the thermoregulatory costs of individuals and prevent frequent energetically costly arousals. Thus, the selection of suitable roosts can have a tremendous impact on energy and water balance of individual bats.

Urban environments provide a large availability of potential roosts but a low diversity of roost types [1]. In cities, many bat species roost in buildings, under bridges, and in other artificial structures. These roosts are usually occupied by individuals for long periods of time, a feature that might represent an advantage for taxa, such as bats, that are at the slow end of the life history continuum [i.e. long-lived, with low reproductive outputs; 3]. Buildings can also serve as roosts for bats that forage in adjacent areas where roosts are scarce [38]. Housing density was one of the main determinants of bat species richness in Sydney, Australia, perhaps due to high roost availability [36]. The benefits of roosting at urban settlements might differ among different groups of bats. For example, in their natural habitats, molossids tend to roost in crevices, which are not very abundant; in principle, the transition from roosting in crevices to roosting in buildings seems very plausible, and these bats will find many potential roost sites in urban habitats [37, 44].

Because artificial roosts in urban environments are usually more sheltered than natural roosts, they may provide higher thermal stability for bats, buffering against external environmental conditions. Indeed, roost type can significantly influence energy expenditure in bats. Czenze et al. [52] found that bats in buffered roosts have a lower heat tolerance compared to those living in roosts that are more exposed, with larger fluctuations in temperature. Marroquin and Muñoz-Garcia (pers. comm.) tested the effect of roost type on mass-specific metabolic rate (msMR, calculated as MR/body mass) of bats while controlling for phylogeny. They gathered data from the literature on msMR of euthermic individuals at ambient temperatures (T_a) ranging from 0 °C to >40 °C (the range of natural roost temperatures) for 46 bat species. They classified roosts into seven categories: (1) "no roost", for exposed surfaces; (2) "tree foliage and bark", which included plant tents, tree bark and foliage, and leaf litter; (3) "rock crevices"; (4) "tree cavities and hollow trees"; (5) "anthropogenic tunnels and mines"; (6) "caves"; and (7) "buildings". To account for the influence of diet on metabolic rates, they assigned bats to one of the five dietary categories: carnivory, frugivory (includes nectarivory), insectivory (mainly consuming arthropods), sanguinivory, and omnivory (consuming plants and animals). For each bat species, they analysed the relationship between msMR and T_a in three temperature regions: (1) below the lower critical temperature, where they estimated conductance (the slope of the relationship between msMR and T_a) for each species; (2) the thermoneutral zone, using the lowest msMR value for each bat species at $T_a \geq 25$ °C (msMR$_{min}$) as a proxy for basal metabolic rate; and (3) above the upper critical temperature, considering values of msMR for 15 species at $T_a = 40$ °C (msMR$_{HT}$).

In their analyses of msMR, Marroquin and Muñoz-Garcia (pers. comm.) found that below the lower critical temperature, conductance was higher for species in more sheltered roosts (buildings and caves) than it was for bats in more exposed roosts (no roost, tree foliage and bark). Also, species in sheltered roosts had a lower msMR$_{min}$ but higher msMR$_{HT}$ than species in exposed roosts. Finally, bats that use more buffered roosts have a higher msMR$_{HT}$ than bats in exposed roosts (Table 1.1, Fig. 1.1). Conductance was higher for insectivores and omnivores than for carnivorous, frugivorous, and sanguivores. Frugivores had the highest msMR$_{min}$, whereas insectivores had the lowest msMR$_{min}$ (Fig. 1.1).

Overall, these results suggest that there is an association between energy expenditure and roost selection in bats. It seems that species that show low conductance at ambient temperatures below the lower critical temperature use more exposed roosts, whereas species with high conductance use sheltered roosts. We suggest that

Table 1.1 Relationship between msMR and T_a below the lower critical temperature in species of bats in different dietary categories and different kinds of roost types

Roost type	Slope of msMR against T_a (mW/g °C)	Intercept of msMR against T_a (mW/g)	Minimum MR (mW/g)	msMR at 40 °C (mW/g)
Buildings	−1.0362	43.66	7.98	11.97 ($n = 3$)
Caves	−1.1478	47.79	9.98	19.26 ($n = 3$)
Man-made tunnels and mines	–	–	8.14 ($n = 2$)	16.05 ($n = 2$)
Rock crevices	−0.419	19.98	4.58 ($n = 1$)	7.21 ($n = 2$)
Tree cavities and hollow trees	−1.0772	37.69	7.41	8.74 ($n = 1$)
Tree foliage and bark	−0.9093	37.92	9.60	9.90 ($n = 3$)
No roost	−0.7174	29.56	6.5 ($n = 1$)	10.51 ($n = 1$)

Data from Marroquin and Muñoz-Garcia (Submitted)

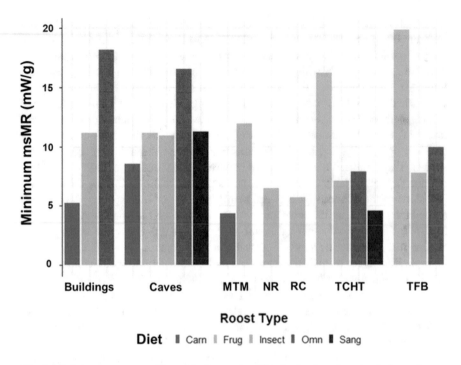

Fig. 1.1 Minimum mass-specific metabolic rate (msMR$_{min}$) of euthermic animals for each roost type. (Data from Marroquin and Muñoz-Garcia, submitted)

bats that live in more exposed roosts may have evolved physiological responses to exposure to low temperatures. For example, decreasing conductance would result in lower msMR at low ambient temperatures. Compared to bats with low conductance, bats that exhibit higher conductance would be forced to use more energy at low temperatures if they roost in exposed locations. We propose that these species evolved behavioural responses; instead of adjusting their metabolic machinery to face low temperatures, they select appropriate roost microclimates in sheltered roosts, such as buildings, to save energy.

In natural habitats, roosts most likely represent valuable and scarce resources for bats, whereas urban environments offer a greater availability of more sheltered roosts, and species that use them likely do not face intense competition for those kinds of roosts [1]. Cities also offer relatively exposed roosts, such as tree cavities, trees, foliage, and others. Bats can also occupy these roosts, although there is evidence that they may often be outcompeted by birds and insects [53]. Therefore, we suggest that most bat species in urban habitats would use sheltered roosts in their natural habitats, which would facilitate the transition to human-made structures.

The urban heat island effect, defined as the phenomenon whereby cities are warmer than outlying areas [54], might also impact energy expenditure in roosts. The higher conductance and the lower msMR$_{min}$ found in building-roosting bats indicate that these bats use less energy at moderate T_as compared to bats in more exposed sites (Marroquin and Muñoz-Garcia, pers. comm.). Thus, we suggest that bats whose natural roosts are more sheltered might be preadapted to roosting in buildings or other anthropogenic structures, which are abundant and readily available in urban environments. The finding that insectivory was the dietary category associated with the highest conductance and the lowest $msMR_{min}$ is consistent with the idea that insectivory is another preadaptation to urban environments [2, 50].

5.2   Water Balance

Bats gain water by drinking and from their food (preformed water). Nonreproductive bats lose water in faeces and urine, but up to 80–85% of water loss is total evaporative water loss [TEWL; 50]. We collected data on TEWL across a wide range of T_a from 22 species of bats and assigned bats to the same, seven abovementioned roost categories (see Roosts and Energetics). Preliminary analysis showed that bats that used exposed roosts had a lower mass-specific TEWL compared to bats that used sheltered roosts, at T_a, ranging from 0 to 45 °C ($P < 0.04$; Fig. 1.2). These results support the hypothesis that bats in exposed roosts mainly rely on physiological adaptations to maintain their water and energy balance. They have low msMR

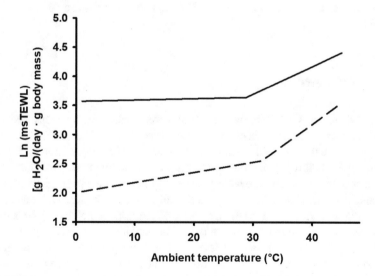

Fig. 1.2 Relationship between mass-specific total evaporative water loss (msTEWL) and ambient temperature in euthermic bats that use exposed roosts (no roost, tree bark and foliage, and rock crevices, dashed line) or sheltered roosts (buildings, caves, man-made tunnels, and tree cavities and hollow trees, solid line). TEWL increases slowly at low and moderate ambient temperatures. Beyond a breakpoint temperature, TEWL increases exponentially. The breakpoint temperature was 31.4 °C for bats in exposed roosts and 28.8 °C for bats in sheltered roosts; this difference was not significant ($P = 0.14$). However, msTEWL was significantly lower at all temperatures in bats that used exposed roosts compared with those bats that used sheltered roosts ($P < 0.04$)

(Table 1.1) and seem to more efficiently use water at high temperatures. In contrast, bats with higher msMR and TEWL rates at high temperatures may mostly depend on behavioural selection of cooler roost microclimates found in buffered roosts, such as buildings.

Urban environments usually have abundant permanent water, which not only provides necessary drinking water for bats with high mass-specific TEWL rates but also increases food availability compared with non-urban areas. For example, *Myotis* spp. exhibit high activity close to water sources in urban environments, where they can also find the water-emergent insects on which they feed [55]. Access to drinking water is often the main limiting resource that determines distribution of bats in their natural habitats [56, 57]. However, water balance of urban bats will likely be less affected than in natural habitats, due to the high availability of drinking water in urban environments.

As in all mammals, food and water requirements of female insectivorous bats increase substantially during pregnancy and even more so during nursing. Milk production by lactating females represents another significant avenue of water loss [58]. Pregnant and lactating females must use warmer roosts to prevent entering torpor, which prolongs gestation and decreases milk production [59]. Warm roosts are relatively abundant in urban environments. However, exposure to higher temperatures also leads to increased evaporative water loss; this fact, combined with the increased water demands of milk production, implies that reproductive females need to drink more often and maternity roosts should be close to water bodies [56, 58].

6   Conclusions

Urban environments pose various insurmountable challenges for many species of bats, which results in their absence from these ecosystems. It is difficult to predict which species will succeed in cities. Urban species have unique strategies to cope with pollution, anthropogenic noise and light, and other environmental factors that are absent (or, at least, less prevalent) in their natural habitats [51]. The findings on birds and mammals and preliminary studies on bats that we present in this chapter highlight how physiological stress responses to anthropogenic disturbances can be modulated by numerous extrinsic and intrinsic factors, which may mask general patterns and should be considered when studying stress physiology. Specific factors that should be considered for bats include their ability to fly, high degree of sociality, trophic niche, and use of diverse roost types. Furthermore, anthropogenic disturbances may have short- or long-term influences on wildlife, and we predict that urban bats show a high degree of phenotypic plasticity in their physiological responses to stress.

Urban bat species seem to share some key traits that make them preadapted to succeed in cities. Species that flourish in urban environments have morphologies that help them fly fast and commute over long distances at a relatively low energetic cost and locate and exploit patchily distributed resources. The abundance of food patches in cities might significantly decrease foraging time and cost and widen trophic niche breadths for bats.

Bats that use sheltered roosts seem to have lower metabolic rates at moderate T_a; because cities are heat islands and are more thermally stable compared to adjacent, non-urban areas, these species of bats might have lower-energy expenditures than their counterparts that use more exposed roosts. Bats that use sheltered roosts seem to have higher mass-specific TEWL rates than those using more exposed roosts, but that will presumably not affect water balance of urban bats due to the high availability of drinking water in these environments.

Overall, there seem to be species of Chiropterans that show an "urban physiological syndrome" that allows them to successfully colonise cities. Urban bats typically are physiologically and behaviourally resistant to urban stressors; are insectivorous or frugivorous; are fast, efficient flyers; use sheltered roosts; and have relatively low metabolic rates compared with other bat species. This syndrome may represent a preadaptation that results in high foraging efficiencies and low energetic costs in urban environments.

Acknowledgements   This is publication 1114 of the Mitrani Department of Desert Ecology.

Literature Cited

1. Russo D, Ancillotto L (2015) Sensitivity of bats to urbanization: a review. Mammal Biol 80(3):205–212. https://doi.org/10.1016/j.mambio.2014.10.003
2. Jung K, Threlfall CG (2018) Trait-dependent tolerance of bats to urbanization: a global meta-analysis. Proc R Soc B 285(1885):20181222. https://doi.org/10.1098/rspb.2018.1222
3. Voigt CC, Kingston T (2016) Bats in the Anthropocene: conservation of bats in a changing world. Springer Nature
4. Moberg GP (1985) Biological response to stress: key to assessment of animal well-being? In: Animal stress. Springer, New York
5. Martin LB (2009) Stress and immunity in wild vertebrates: timing is everything. Gen Comp Endocrinol 163(1–2):70–76. https://doi.org/10.1016/j.ygcen.2009.03.008
6. Dantzer B, Fletcher QE, Boonstra R, Sheriff MJ (2014) Measures of physiological stress: a transparent or opaque window into the status, management and conservation of species? Conserv Physiol 2(1):cou023. https://doi.org/10.1093/conphys/cou023
7. Wingfield JC, Smith JP, Farner DS (1982) Endocrine responses of white-crowned sparrows to environmental stress. Condor 84(4):399–409. https://doi.org/10.2307/1367443
8. Touma C, Palme R, Sachser N (2004) Analyzing corticosterone metabolites in fecal samples of mice: a noninvasive technique to monitor stress hormones. Horm Behav 45(1):10–22. https://doi.org/10.1016/j.yhbeh.2003.07.002
9. Beerda B, Schilder MB, Janssen NS, Mol JA (1996) The use of saliva cortisol, urinary cortisol, and catecholamine measurements for a noninvasive assessment of stress responses in dogs. Horm Behav 30(3):272–279. https://doi.org/10.1006/hbeh.1996.0033
10. Davenport MD, Tiefenbacher S, Lutz CK, Novak MA, Meyer JS (2006) Analysis of endogenous cortisol concentrations in the hair of rhesus macaques. Gen Comp Endocrin 147(3):255–261. https://doi.org/10.1016/j.ygcen.2006.01.005

11. Romero LM (2002) Seasonal changes in plasma glucocorticoid concentrations in free-living vertebrates. Gen Comp Endocr 128(1):1–24. https://doi.org/10.1016/S0016-6480(02)00064-3

12. Beaugeard E, Brischoux F, Henry PY, Parenteau C, Trouvé C, Angelier F (2019) Does urbanization cause stress in wild birds during development? Insights from feather corticosterone levels in juvenile house sparrows (*Passer domesticus*). Ecol Evol 9(1):640–652. https://doi.org/10.1002/ece3.4788

13. Bonier F (2012) Hormones in the city: endocrine ecology of urban birds. Horm Behav 61(5):763–772. https://doi.org/10.1016/j.yhbeh.2012.03.016

14. Iglesias-Carrasco M, Aich U, Jennions MD (1936) Head ML (2020) stress in the city: meta-analysis indicates no overall evidence for stress in urban vertebrates. Proc R Soc B 287:20201754. https://doi.org/10.1098/rspb.2020.1754

15. Allen LC, Turmelle AS, Widmaier EP, Hristov NI, Mccracken GF, Kunz TH (2011) Variation in physiological stress between bridge- and cave-roosting Brazilian free-tailed bats. Conserv Biol 25(2):374–381. https://doi.org/10.1111/j.1523-1739.2010.01624.x

16. Ditchkoff SS, Saalfeld ST, Gibson CJ (2006) Animal behavior in urban ecosystems: modifications due to human-induced stress. Urban Ecosyst 9(1):5–12. https://doi.org/10.1007/s11252-006-3262-3

17. Parry-Jones K, Webster KN, Divljan A (2016) Baseline levels of faecal glucocorticoid metabolites and indications of chronic stress in the vulnerable grey-headed flying-fox, *Pteropus poliocephalus*. Aust Mammal 38(2):195–203. https://doi.org/10.1071/AM15030

18. Wada H, Yates DE, Evers DC, Taylor RJ, Hopkins WA (2010) Tissue mercury concentrations and adrenocortical responses of female big brown bats (*Eptesicus fuscus*) near a contaminated river. Ecotoxicology 19(7):1277–1284. https://doi.org/10.1007/s10646-010-0513-0

19. Medina-Cruz GE, Salame-Méndez A, Briones-Salas M (2020) Glucocorticoid profiles in frugivorous bats on wind farms in the Mexican tropics. Acta Chiropterol 22(1):147–155. https://doi.org/10.3161/15081109ACC2020.22.1.013

20. Falchi F, Cinzano P, Duriscoe D, Kyba CC, Elvidge CD, Baugh K et al (2016) The new world atlas of artificial night sky brightness. Sci Adv 2(6):e1600377. https://doi.org/10.1126/sciadv.1600377

21. Stone EL, Jones G, Harris S (2009) Street lighting disturbs commuting bats. Curr Biol 19(13):1123–1127. https://doi.org/10.1016/j.cub.2009.05.058

22. Russo D, Ancillotto L, Cistrone L, Libralato N, Domer A, Cohen S, Korine C (2019) Effects of artificial illumination on drinking bats: a field test in forest and desert habitats. An Conserv 22(2):124–133. https://doi.org/10.1111/acv.12443

23. Khan ZA, Yumnamcha T, Mondal G, Devi SD, Rajiv C, Labala RK, Sanjita D, Chattoraj A (2020) Artificial Light At Night (ALAN): a potential anthropogenic component for the COVID-19 and HCoVs outbreak. Front Endocrinol 10:622. https://doi.org/10.3389/fendo.2020.00622

24. Emmer KM, Russart KL, Walker WH II, Nelson RJ, DeVries AC (2018) Effects of light at night on laboratory animals and research outcomes. Behav Neurosci 132(4):302–314. https://doi.org/10.1037/bne0000252

25. Ouyang JQ, de Jong M, Hau M, Visser ME, van Grunsven RH, Spoelstra K (2015) Stressful colours: corticosterone concentrations in a free-living songbird vary with the spectral composition of experimental illumination. Biol Lett 11(8):20150517. https://doi.org/10.1098/rsbl.2015.0517

26. Injaian AS, Francis CD, Ouyang JQ, Dominoni DM, Donald JW, Fuxjager MJ et al (2020) Baseline and stress-induced corticosterone levels across birds and reptiles do not reflect urbanization levels. Conserv Physiol 8(1):coz110. https://doi.org/10.1093/conphys/coz110

27. Fonken LK, Workman JL, Walton JC, Weil ZM, Morris JS, Haim A, Nelson RJ (2010) Light at night increases body mass by shifting the time of food intake. Pro Natl Acad Sci 107(43):18664–18669. https://doi.org/10.1073/pnas.1008734107

28. Hoffmann J, Palme R, Eccard JA (2018) Long-term dim light during nighttime changes activity patterns and space use in experimental small mammal populations. Environ Pollut 238:844–851. https://doi.org/10.1016/j.envpol.2018.03.107
29. Shannon G, McKenna MF, Angeloni LM, Crooks KR, Fristrup KM, Brown E, Warner KA, Nelson MD, White C, Briggs J, McFarland S (2016) A synthesis of two decades of research documenting the effects of noise on wildlife. Biol Rev 91(4):982–1005. https://doi.org/10.1111/brv.12207
30. Navarro-Castilla Á, Mata C, Ruiz-Capillas P, Palme R, Malo JE, Barja I (2014) Are motorways potential stressors of roadside wood mice (*Apodemus sylvaticus*) populations? PLoS One 9:e91942. https://doi.org/10.1371/journal.pone.0091942
31. Brearley G, McAlpine C, Bell S, Bradley A (2012) Influence of urban edges on stress in an arboreal mammal: a case study of squirrel gliders in Southeast Queensland. Australia Landscape Ecol 27(10):1407–1419. https://doi.org/10.1007/s10980-012-9790-8
32. Łopucki R, Klich D, Ścibior A, Gołębiowska D (2019) Hormonal adjustments to urban conditions: stress hormone levels in urban and rural populations of *Apodemus agrarius*. Urban Ecosyst 22(3):435–442. https://doi.org/10.1007/s11252-019-0832-8
33. Domer A, Korine C, Slack M, Rojas I, Mathieu D, Mayo A, Russo D (2021) Adverse effects of noise pollution on foraging and drinking behaviour of insectivorous desert bats. Mammal Biol 101(4):497–501. https://doi.org/10.1007/s42991-021-00101-w
34. Schaub A, Ostwald J, Siemers BM (2009) Foraging bats avoid noise. J Exp Biol 212(19):3174–3180. https://doi.org/10.1242/jeb.022863
35. Bunkley JP, Barber JR (2015) Noise reduces foraging efficiency in pallid bats (*Antrozous pallidus*). Ethology 121(1):1116–1121. https://doi.org/10.1111/eth.12428
36. Threlfall C, Law B, Penman T, Banks PB (2011) Ecological processes in urban landscapes: mechanisms influencing the distribution and activity of insectivorous bats. Ecography 34(5):814–826. https://doi.org/10.1111/j.1600-0587.2010.06939.x
37. Jung K, Kalko EKV (2011) Adaptability and vulnerability of high flying Neotropical aerial insectivorous bats to urbanization. Divers Distrib 17(2):262–274. https://doi.org/10.1111/j.1472-4642.2010.00738.x
38. Duchamp JE, Swihart RK (2008) Shifts in bat community structure related to evolved traits and features of human-altered landscapes. Landsc Ecol 23:849–860. https://doi.org/10.1007/s10980-008-9241-8
39. Jung K, Kalko EKV (2010) Where forest meets urbanization: foraging plasticity of aerial insectivorous bats in an anthropogenically altered environment. J Mammal 91(1):144–153. https://doi.org/10.1644/08-MAMM-A-313R.1
40. Kelm DH, Popa-Lisseanu AG, Dehnhard M, Ibáñez C (2016) Non-invasive monitoring of stress hormones in the bat *Eptesicus isabellinus*–do fecal glucocorticoid metabolite concentrations correlate with survival? Gen Comp Endocr 226:27–35. https://doi.org/10.1016/j.ygcen.2015.12.003
41. Reeder DM, Kosteczko NS, Kunz TH, Widmaier EP (2006) The hormonal and behavioral response to group formation, seasonal changes, and restraint stress in the highly social Malayan flying fox (*Pteropus vampyrus*) and the less social little golden-mantled flying fox (*Pteropus pumilus*) (Chiroptera: Pteropodidae). Horm Behav 49(4):484–500. https://doi.org/10.1016/j.yhbeh.2005.11.001
42. Lewanzik D, Kelm DH, Greiner S, Dehnhard M, Voigt CC (2012) Ecological correlates of cortisol levels in two bat species with contrasting feeding habits. Gen Comp Endocr 177(1):104–112. https://doi.org/10.1016/j.ygcen.2012.02.021
43. Geggie JF, Fenton MB (1985) A comparison of foraging by *Eptesicus fuscus* (Chiroptera: Verspertilionidae) in urban and rural environments. Can J Zool 63(2):263–266. https://doi.org/10.1139/z85-040
44. Avila-Flores R, Fenton BM (2005) Use of spatial features by foraging insectivorous bats in a large urban landscape. J Mammal 86(6):1193–1204. https://doi.org/10.1644/04-MAMM-A-085R1.1

45. Pellón JJ, Mendoza JL, Quispe-Hure O, Condo F, Williams M (2021) Exotic cultivated plants in in the diet of the nectar-feeding bat *Glossophaga soricina* (Phyllostomidae: Glossophaginae) in the city of Lima. Peru Acta Chiropterol 23(1):107–117. https://doi.org/10.3161/15081109 ACC2021.23.1.009

46. Popa-Lisseanu AG, Bontadina F, Ibáñez C (2009) Giant noctule bats face conflicting constraints between roosting and foraging in a fragmented and heterogenous landscape. J Zool 278(2):126–133. https://doi.org/10.1111/j.1469-7998.2009.00556.x

47. Acre M, Michopoulos V, Shepard KN, Ha QC, Wilson ME (2010) Diet choice, cortisol reactivity, and emotional feeding in socially housed rhesus monkeys. Physiol Behav 101(4):446–455. https://doi.org/10.1016/j.physbeh.2010.07.010

48. Kitaysky AS, Wingfield JC, Piatt JF (1999) Dynamics of food availability, body condition and physiological stress response in breeding black-legged kittiwakes. Funct Ecol 13(5):577–584. https://doi.org/10.1046/j.1365-2435.1999.00352.x

49. Miguel PH, Kerches-Rogeri P, Niebuhr BB, Cruz RAS, Ribeiro MC, da Cruz Neto AP (2019) Habitat amount partially affects physiological condition and stress level in Neotropical fruit-eating bats. Comp Biochem Physiol A: Molecular & Integrative Physiology 237:110537. https://doi.org/10.1016/j.cbpa.2019.110537

50. Neuweiler G (2000) The biology of bats. Oxford University Press, Oxford and New York

51. Kunz TH, Lundsem LF (2003) Ecology of cavity and foliage roosting bats. In: Kunz TH, Fenton MB (eds) Bat ecology. The University of Chicago Press, Chicago, pp 430–490

52. Czenze ZJ, Smit B, van Jaarsveld B, Freeman MT, McKechnie AE (2021) Caves, crevices and cooling capacity: roost microclimate predicts heat tolerance in bats. Funct Ecol. https://doi.org/10.1111/1365-2435.13918

53. Threlfall CG, Law B, Banks PB (2013) Roost selection in suburban bushland by the urban sensitive bat *Nyctophilus gouldi*. J Mammal 94(2):307–319. https://doi.org/10.1644/11-MAMM-A-393.1

54. Shochat E, Warren PS, Faeth SH, McIntyre NE, Hope D (2006) From patterns to emerging processes in mechanistic urban ecology. Trends Ecol Evol 21(4):186–191. https://doi.org/10.1016/j.tree.2005.11.019

55. Dixon MD (2012) Relationship between land cover and insectivorous bat activity in an urban landscape. Urban Ecosyst 15(3):683–695. https://doi.org/10.1007/s11252-011-0219-y

56. Korine C, Daniel S, Pinshow B (2013) Roost selection by female Hemprich's long-eared bats. Behav Process 100:131–138. https://doi.org/10.1016/j.beproc.2013.08.013

57. Cappelli MP, Blakey RV, Taylor D, Flanders J, Badeen T, Butts S, Frick WF, Rebelo H (2021) Limited refugia and high velocity range-shifts predicted for bat communities in drought-risk areas of the Northern Hemisphere. Glob Ecol Conserv 28:e01608. https://doi.org/10.1016/j.gecco.2021.e01608

58. Adams RA, Hayes MA (2008) Water availability and successful lactation by bats as related to climate change in arid regions of western North America. J An Ecol 77(6):1115–1121. https://doi.org/10.1111/j.1365-2656.2008.01447.x

59. Wilde CJ, Knight CH, Racey PA (1999) Influence of torpor on milk protein composition and secretion in lactating bats. J Exp Zool 284(1):35–41. https://doi.org/10.1002/(SICI)1097-010X(19990615)284:1<35:AID-JEZ6>3.0.CO;2-Z

Chapter 2
Genetic Impoverishment in the Anthropocene: A Tale from Bats

Balaji Chattopadhyay, Kritika M. Garg, Rajasri Ray, Ian H. Mendenhall, and Frank E. Rheindt

Abstract Habitat loss, fragmentation, and anthropogenic climate change are major drivers of biodiversity declines during the ongoing Anthropocene epoch. Understanding the evolutionary trajectories of organisms with diverse life histories in response to these threats can enable us to predict the fate of the extant biota facing accelerated habitat loss and climate change. Genetic data contain vital clues about species diversity and have been widely used to assess the impacts of non-anthropogenic climate change (since the Last Glacial Maximum and during the Holocene) on a range of species. Recent advances in sequencing technologies and analytical approaches have broadened the scope of genetic investigations. They have allowed us to directly test for recent population bottlenecks linked to rapid, anthropogenic environmental change. In this chapter, we discuss the utility of genomic data in identifying evolutionary trajectories of bats in response to climate change and habitat modification. We show that these nocturnal mammals are particularly sensitive to environmental and habitat fluctuations. We also summarise and discuss our recent investigations of an urban population of the lesser short-nosed fruit bat (*Cynopterus brachyotis*) from the island nation of Singapore and assess the

B. Chattopadhyay (✉)
Trivedi School of Biosciences, Ashoka University, Sonipat, Haryana, India

Department of Biology, Ashoka University, Sonipat, Haryana, India
e-mail: balaji.chattopadhyay@ashoka.edu.in

K. M. Garg
Department of Biology, Ashoka University, Sonipat, Haryana, India

Centre for Interdisciplinary Archaeological Research, Ashoka University, Sonipat, India

R. Ray
Centre for Studies in Ethnobiology, Biodiversity and Sustainability (CEiBa),
West Bengal, India

I. H. Mendenhall
Duke-NUS Medical School, Singapore, Singapore

F. E. Rheindt
Department of Biological Sciences, National University of Singapore, Singapore, Singapore

response of this population to rapid urbanisation during the Anthropocene. Comparisons of genetic diversity estimates and evolutionary models through coalescent simulations revealed that this local population had been on a slow decline for centuries but faced a more drastic bottleneck a few decades ago. We also observed an astonishing level of decline in indicators of genetic diversity in the local population over the past century, coinciding with the rapid urbanisation of Singapore. Our observations show that even commonly occurring, synanthropic species of bat (i.e. lives within and appears to benefit from urban environments) have been negatively impacted by rapid urbanisation. Our results also highlight the necessity of assessing the impact of urban green spaces on the evolution and survival of organisms like bats, which often rely on these remnant habitats.

Keywords *Cynopterus brachyotis* · Bottleneck · Genomic diversity · Lesser short-nosed fruit bat · Urbanisation

1 Introduction

The Anthropocene is the age of human dominance and urbanisation [1–4], characterised by rapid loss of natural habitats, climate change, and an ongoing biotic extinction crisis of unprecedented proportions [1, 2, 4, 5]. Recent human activities associated with global industrialisation have radically altered ecosystems, with effects approaching the magnitude of natural events such as tectonic movements, glacial cycles, volcanic eruptions, and other major disruptions [4]. The precise start date of the Anthropocene epoch is still debated, but for the purposes of this chapter, we set it at 1945, after which humanity has undergone a period of 'great acceleration' during which human activities increased on an exponential scale [1]. These activities drove an almost unparalleled episode of destruction of natural habitats, sudden rise in global temperature, alteration of the prevalent climate, and the sixth mass extinction in the history of life [2]. The Earth is being stripped of its biodiversity as species are lost at a rate that is magnitudes higher than that of previous mass extinctions, in which comparable losses occurred across tens of millions of years [2, 6]. We are losing species that have not even been discovered or described [7], prompting scientists and governments to prioritise efforts to understand and mitigate damages to wildlife.

One of the key goals of contemporary biological sciences is to understand the vulnerability of species or biotic communities to habitat modification and climate change, not only to conserve and manage biodiversity but also to understand the impact of biodiversity loss on human survival and well-being. Species' vulnerability to future climate change can sometimes be predicted based on their responses to historic climatic alterations and concomitant habitat fluctuations [8, 9]. Quaternary climatic fluctuations caused multiple periods of glacial maxima over the past two million years that resulted in repeated range shifts for many species globally [8–11]. Significant reduction in global temperatures led to increased glaciation and aridity. This reduced the available habitat for many taxa and forced them into isolated, local refugia. However, sea levels also fell during these periods, connecting previously isolated landmasses (through land bridges) and expanding available habitat for many other taxa [10–12]. These alternating periods of high isolation and high connectivity drove fluctuations in population size and genetic diversity and in some cases increased rates of speciation [8–11, 13, 14]. For example, increasing subdivision and speciation caused by Pleistocene glacial cycles are documented in many bird species that inhabit islands and prefer forested habitats (specialists), while bird species that use a much wider range of habitats (habitat generalists) exhibit lower rates of speciation [13].

Bats are an interesting group with which to study the effects of climate change on population size and genetic diversity. The order Chiroptera is the second largest mammalian order, comprising more than 1400 known species of bat and harbouring an enormous range of cryptic diversity [15–19]. Bats are also keystone species to many ecosystems and serve as excellent bioindicators [20–22]. They are very sensitive to abrupt climatic fluctuations, and heat stress during heatwaves has been linked to mass mortalities in several regions [23–25]. Large, frugivorous species are susceptible to heatwaves, as shown by mass mortalities of fruit bats recorded in Australia during summer heatwaves [25]. The impact of climate change on bats is not always negative as it may enable range expansions to cooler regions as observed in Mexican free-tailed bats (*Tadarida brasiliensis*) [26]. Bats are ubiquitous and often associated with human-modified habitat [20–22], but their evolutionary response to climate change, both historical and current (i.e. human-mediated), is unclear.

To fill this lacuna in the framework of a long-term initiative to understand the effects of climate change on wildlife, we assessed how bats have responded to historic climatic fluctuations and further investigated the links between urbanisation and population endangerment. While studying the long-term effects of Earth's historic climate change, we reconstructed species' evolutionary histories from single genomes (PSMC analysis: pairwise sequentially Markovian coalescent) and palaeohabitats of 11 phylogenetically divergent bat species with a wide range of biological and ecological traits [8]. In that study, we assessed whether changes in palaeohabitats during periods of climatic fluctuations correlate with changes in effective population size. We observed a significant correlation between available palaeohabitat and effective population size during the last glacial period. Frugivores were particularly susceptible to global warming, with their population size dramatically decreasing after the Last Glacial Maximum. Our comparative genomic analysis also indicated that large insectivores generally have a low effective population size and that bat species generally entered the Holocene with low effective population sizes [8]. These observations indicate overall vulnerability of bats to climate change and concomitant habitat fluctuations. They also suggest that a species' biology and ecology play an important role in determining its resilience during future climate change and habitat modifications [8]. Comparative PSMC analyses across vertebrates remain rare in the scientific literature. However, the available studies (e.g. 6 felids and 38 species of birds) reveal similar patterns of declines in effective population size during the last glacial period for most taxa and especially those that are currently endangered [9, 27].

Studying species' responses to historic climatic fluctuations has highlighted that knowledge of the past can shed light on the effects of more recent climate change. This includes the changes during the past few centuries marked by the industrial revolution in general and specifically during the last few decades that were characterised by ultra-rapid urbanisation, mechanisation, and habitat loss [28]. In many cases, however, the available methods of demographic reconstruction cannot sufficiently capture the signals of these recent effects [28–30]. Rather, a comparison of pre- and post-decline populations can alleviate this limitation and provide resolution of demographic histories during the Anthropocene [28, 30]. In this context, biological collections are a treasure trove. Specimens collected over the last few centuries and preserved in natural history museums worldwide can provide us with genetic samples that reflect a population's status prior to intense urbanisation. This enables a direct comparison with the contemporary (post-urbanisation) population and generates a much deeper resolution into the nature of demographic fluctuations [30]. Next-generation sequencing and bioinformatic pipelines capable of analysing large datasets make it possible to generate and analyse genomic data from degraded museum specimens. This approach has provided insights into historical factors associated with endangerment of natural populations [28, 31, 32]. Armed with these technological advances, scientists hope to leverage species histories to predict their responses to environmental change and their mid- and long-term viability [8, 28].

In old museum specimens, DNA is often heavily degraded due to storage time and chemical damage from historically popular preservatives such as formaldehyde [33]. This has long limited the utility of this valuable resource for genomic analyses. Propitiously, it is now possible to generate large-scale DNA sequencing data even from degraded biological material preserved in museums [33]. For example, a recent study compared DNA from a woolly mammoth (*Mammuthus primigenius*) fossil from about 45,000 years ago, when mammoths were plentiful across the Holarctic region and the effective population size was around 13,000, with those retrieved from a sample dating 4300 years ago from Wrangel Island in the Arctic Ocean, where the population consisted of roughly 300 individuals and represented one of the species' last strongholds prior to extinction [34]. The analysis revealed the accumulation of detrimental mutations in the isolated, island population, consistent with the hypothesis of 'genomic meltdown' prior to extinction.

2   Temporal Genomic Data Reveal Drastic Population Genetic Diversity Decline in a Tropical Fruit Bat

Although bats are one of the most common mammals in anthropogenically altered habitats, the effects of urbanisation on the genetic diversity of bat populations are unclear [8, 22, 28]. As part of our research to understand the sensitivity of bats to Anthropocene habitat alterations, we investigated potential genomic impoverishment in the Sunda lineage of the lesser short-nosed fruit bat (*Cynopterus brachyotis*) following urbanisation of the island nation of Singapore. This medium-sized, generalist fruit bat is widely distributed across Southeast Asia [35, 36]. It is synanthropic (lives within and appears to benefit from urban environments) and is often observed in cities, towns, and villages, in close proximity to humans. *Cynopterus brachyotis* is broadly sympatric with at least two congeners: the larger Horsfield's fruit bat (*C. horsfieldii*) and a smaller, forest-dwelling species whose taxonomic affinity remains unresolved [36].

Based on its synanthropic tendencies, we predicted that *C. brachyotis* would be resilient to deforestation and would be able to use clusters of urban green space for foraging. The Singapore Strait has largely isolated the population of *C. brachyotis* of Singapore from its nearest neighbours in Peninsular Malaysia since the beginning of the Holocene [37, 38]. Therefore, this study location allowed us to confidently exclude most of the confounding effects of migration and gene flow.

Over the past century, Singapore lost about 95% of its forest cover in parallel massive urbanisation and industrialisation. This caused a considerable loss in biodiversity and an estimated 34–87% loss of species in some taxa experiencing local extinctions [7, 39]. However, careful planning and management have facilitated tree canopy cover over 30% of the island's area, which is high compared with many other large cities (http://senseable.mit.edu/treepedia/cities/singapore).

To understand the impact of urbanisation on a commonly occurring species, we compared the genomic diversity in *C. brachyotis* collected before and after the intensive urbanisation of Singapore. We sampled bats collected in 1931 (pre-industrialisation, pre-urbanisation; $n = 21$) from the Lee Kong Chian Natural History Museum (LKCNHM), Singapore [28], and from the contemporary population, sampled in 2011–2012 ($n = 20$). Genomic diversity of these historic samples was compared to contemporary samples collected in 2011–2012 ($n = 20$). We isolated DNA in dedicated ancient DNA facilities designed for work with historic samples. We targeted ~1.5 Mb of the *C. brachyotis* genome through sequence capture methods, which are highly effective in comparing similar regions of the genome among samples [28, 40]. We designed our own sequence capture panel to target 1184 loci distributed across the genome (see Fig. 2.1 for target locus design). These loci were

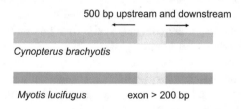

Fig. 2.1 Target locus design used to isolate genome-wide data (see Chattopadhyay et al. [28] for further details). In brief, we first selected exons from the little brown bat (*Myotis lucifugus*) genome that are more than 200 bp long and are conserved across bats. Following this, we identified these exons in the in-house generated genome of the lesser short-nosed fruit bat (*Cynopterus brachyotis*). For every such exon in *C. brachyotis* genome, we also retrieved sequence data 500 bp upstream and downstream; each of these intron-exon-intron segments was considered one locus for the study

amplified using sequence capture protocols and sequenced on an Illumina HiSeq4000 platform (see Chattopadhyay et al. [28] for further details).

We generated over 634 million reads and retained approximately 483 million reads after clean-up (removal of adapters, low-quality reads, PCR duplicates) [28]. Historic samples carried characteristic signatures of DNA damage: excessive cytosine to thymine substitutions at the 5′ end and guanine to adenine substitutions at the 3′ end. All reads from these historic samples were rescaled and trimmed using map-Damage 2 [41]. The cleaned reads were processed in two ways: first, by generating sequence data for all 1184 target loci (using HybPiper pipeline 1.2 [42] and, second, by aligning these reads directly to the *C. brachyotis* genome and mining genome-wide single-nucleotide polymorphisms (SNPs) (ANGSD [43]). This dual approach helped us to generate diversity estimates informed by over a million base pairs of sequence data and thousands of SNP loci. Our analyses required genetic markers that are unlinked (i.e. have independent evolutionary histories), are selectively neutral (i.e. do not directly affect the fitness of the organism and are therefore not under selection) and have not undergone recombination. We checked our data for signatures of these evolutionary processes and pruned loci identified in this screening, retaining 874 loci for sequence-based analysis (990,087 bp) and 24,782 SNPs (see Chattopadhyay et al. [28] for additional details).

2.1 Decline in Genetic Diversity

We compared genetic diversity and inbreeding coefficients of the historic and contemporary populations of *C. brachyotis*. For the sequence data, we estimated the number of variable sites, number of parsimony informative sites, proportion of variable sites, and proportion of parsimony informative sites using AMAS [44]. We observed significant differences in genomic diversity estimated for all four summary statistics, with higher genomic diversity in historic populations than in contemporary ones (Fig 2.2a–d). We used the SNP data to measure internal relatedness, homozygosity by loci, proportion of heterozygous loci, and standardised heterozygosity relative to mean expected heterozygosity, using the R package GENHET [45, 46], pairwise relatedness using COANCESTRY 1.0.1.7 [47], and inbreeding coefficients with PLINK 1.9 [48]. We also generated estimates of genome-wide probability of heterozygosity per individual in ANGSD for all individuals. All SNP-based summary statistics other than homozygosity by loci were significantly different between historic and contemporary populations (Fig. 2.2e–k). All summary statistics suggested that the contemporary population is genetically impoverished relative to the historic sample, with higher inbreeding coefficients and higher pairwise relatedness between individuals (Fig. 2.2).

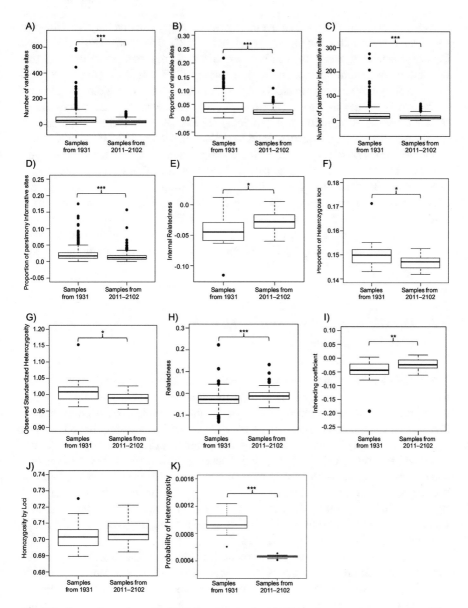

Fig. 2.2 Comparison of genetic diversity estimates for historic and contemporary samples of the lesser short-nosed fruit bat (*Cynopterus brachyotis*) from Singapore based on the sequence dataset (A–D) and SNP dataset (E–K). * denotes *p* values less than 0.05, ** denotes *p* values less than 0.01, and *** denotes *p* values less than 0.001

Our observations illustrate an overall reduction in genetic diversity over the past ~90 years within the Singapore population of *C. brachyotis* (Fig. 2.2). The Palaeotropical ecosystems of Asia have experienced large-scale destruction of forests and other natural habitats in the past century [7, 39, 49–51]. This region also harbours major biodiversity hotspots and cryptic diversity, with ongoing and regular discovery of species new to science [49, 50, 52]. However, the pace and extent of ongoing habitat destruction and the local effects of global climate change in Southeast Asia may drive many species to extinction even before they are described [7, 39, 49–51]. The island nation of Singapore is a microcosm of these challenges, with rapid urbanisation during the past century in conjunction with drastic deforestation driving local extinction of multiple species during the last five decades [7, 39, 53, 54]. *Cynopterus brachyotis* can be synanthropic but is also a keystone species in forests acting as both pollinator and seed disperser [55]. Declining genetic diversity in the Singapore population of *C. brachyotis* suggests population declines concurrent with urbanisation, implying that urbanisation can drive even ubiquitous species to become rare [28].

2.2 Drastic Population Bottleneck Coinciding with Urbanisation

The correlation between low genetic diversity and extinction is well established in the scientific literature; genetic factors partly predict the viability of small populations [56, 57]. Accumulation of deleterious mutations and overall loss of genetic diversity in some isolated populations have also been speculatively linked to local extinctions [31, 34]. Temporal sampling allows the empirical evaluation of genetic diversity loss over time [28–30, 58], and reduced genetic diversity between historic and contemporary samples implies fluctuations in population size. Comparisons of ancient and contemporary diversity have revealed demographic trends over long periods of time (e.g [58].). However, application of temporal sampling to understand the potential effects of the Anthropocene on the genetic diversity of wild populations is a relatively new approach [28, 30, 59, 60].

To understand the effects of human-mediated changes on the evolutionary trajectory of the Singapore population of *C. brachyotis*, we assessed support for competing models of demographic history. We constructed six different scenarios of population decline, including gradual decline, population bottlenecks (i.e. rapid, dramatic decreases in population size), and a combination of both decline and bottlenecks. We also included a model of constant population size and a model for gradual population expansion. We used the site frequency spectrum (SFS) for demographic reconstructions in fastsimcoal 2.6.02 [61] to compare these eight models (see Chattopadhyay et al. [28] for further details). We used a temporal sampling approach for demographic reconstructions as it can robustly identify bottlenecks under most circumstances compared to only analysing modern samples [29].

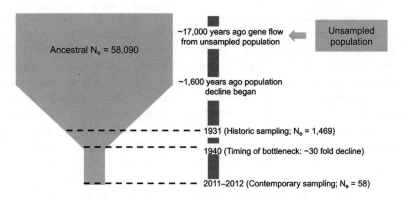

Fig. 2.3 Visual representation of the best fit model of demographic history of the Singapore population of the lesser short-nosed fruit bat (*Cynopterus brachyotis*)

We observed that the best model was the most complex one (Fig. 2.3), positing historic gene flow from an ancestral, unsampled population and a continuous decline that started 195 generations ago (range: 63 to 507 generations ago). Assuming a generation time of 8 years for *C. brachyotis*, this suggests that this decline started approximately 1600 years ago. This model also estimated a very recent bottleneck at nine generations ago (~1939, but range 2–11 generations ago).

2.3   Common Species Are Not Immune to the Effects of Fragmentation

Our demographic models provide strong evidence of direct impacts of urbanisation on natural populations, which may also apply to wildlife in other rapidly changing landscapes. The estimated timing of the recent bottleneck in the Singapore population of *C. brachyotis* coincides with the advent of the Anthropocene [1], a period of rapid urbanisation and drastic decline of forest cover in the island nation [7, 39, 54]. Ongoing efforts to promote habitat regeneration in Singapore may support the recovery of this and other affected populations, and long-term monitoring of these populations can assess the impact of such efforts.

The severity of the estimated bottleneck is considerable, as the nearly 30-fold decrease (range: 3–96-fold decline) is likely detrimental to population viability (Fig. 2.3). Bottlenecks increase a population's vulnerability to stochastic events increasing the probability of local extinctions [56, 57, 62]. Loss of genetic diversity and population bottlenecks due to human interference has been documented in many endangered vertebrates [5, 31, 56]. Our results build on this literature by demonstrating these effects in a common species in response to urbanisation.

The low genetic diversity and effective population size of *C. brachyotis* in Singapore (Fig. 2.3) raise concerns about the long-term viability of this island population and provide a sobering example of the effect that urbanisation may have on wildlife. Human-mediated habitat alterations are not only severely detrimental to the survival of rare species but also affect common species coexisting closely with humans [28]. The Anthropocene, the epoch of human dominance, is synonymous with devastating defaunation due to human activities [1, 2, 4]. A recent study showed declining population sizes and shrinking ranges in ~9000 vertebrates, including common species, over the last century [6]. These ongoing declines imply an accelerated pace of genetic impoverishment that can interact with small population sizes to increase extinction risk [6]. Our study illustrates how modern sequencing technologies and genomic analyses can untangle the patterns and processes driving diversity loss, elucidating its causes and consequences in the medium and long term. Our study also illustrates the value of museum collections in enabling temporal sampling and comparisons of pre- and post-urbanisation populations, to quantify the true effects of the Anthropocene. We particularly acknowledge the value of museum collections that have been built by the massive efforts of various collectors in the past.

3   Conclusion

Taken together, our observations from bats provide insights into the effects of human-induced climate change and landscape modifications, particularly urbanisation, on resident wildlife. In a synthesis of multiple studies, we found that bats are particularly sensitive to climatic fluctuations and habitat loss. Our integrative research paradigm connecting past climate and habitat alterations with genome-scale data allowed us to identify detailed evolutionary patterns of natural populations from present diversity to historical fluctuations going back millions of years. Our results add to the growing body of literature showing the threat posed to wildlife by human encroachment. They also provide direct evidence of genetic impacts of urban development, indicating that even common, synanthropic species are threatened by urbanisation and climate change. Fruit bats such as *C. brachyotis* are keystone species that play a primary role in pollination, germination, and regeneration of native tree species [55, 63]. As such, the decline of these bats may adversely affect not only the function of the ecosystems they inhabit but also their long-term ability to support populations of humans and other wildlife.

4   Acknowledgements

B.C. acknowledges funding from the Trivedi School of Biosciences start-up fund. IHM acknowledges NUS Global Asia Institute Grant NIHA-2011-1-005 for sampling. KMG acknowledges the support of DBT Ramalingaswami Fellowship (No.

BT/HRD/35/02/2006). The authors thank the Lee Kong Chian Natural History Museum for providing historic samples.

5   Data Availability

The data used for this study are available on Sequence Read Archive (project accession number is PRJNA666066) and GenBank (genome accession ID: GCA_009793145.1).

Literature Cited

1. Corlett RT (2015) The Anthropocene concept in ecology and conservation. Trends Ecol Evol 30(1):36–41
2. Dirzo R, Young HS, Galetti M, Ceballos G, Isaac NJ, Collen B (2014) Defaunation in the Anthropocene. Science 345(6195):401–406
3. Lewis SL, Maslin MA (2015) Defining the anthropocene. Nature 519(7542):171–180
4. Laurance WF (2019) The Anthropocene. Curr Biol 29(19):R953–R954
5. Chattopadhyay B, Garg KM, Yun Jing S, Low GW, Frechette J, Rheindt FE (2019) Conservation genomics in the fight to help the recovery of the critically endangered Siamese crocodile *Crocodylus siamensis*. Mol Ecol 28:936–950
6. Ceballos G, Ehrlich PR, Dirzo R (2017) Biological annihilation via the ongoing sixth mass extinction signaled by vertebrate population losses and declines. Proc Natl Acad Sci U S A 114(30):E6089–E6096
7. Sodhi NS, Koh LP, Brook BW, Ng PK (2004) Southeast Asian biodiversity: an impending disaster. Trends Ecol Evol 19(12):654–660
8. Chattopadhyay B, Garg KM, Ray R, Rheindt FE (2009) Fluctuating fortunes: genomes and habitat reconstructions reveal global climate-mediated changes in bats' genetic diversity. Proc Royal Soc B 286(1911):20190304
9. Nadachowska-Brzyska K, Li C, Smeds L, Zhang G, Ellegren H (2015) Temporal dynamics of avian populations during Pleistocene revealed by whole-genome sequences. Curr Biol 25(10):1375–1380
10. Hewitt GM (2000) The genetic legacy of the quaternary ice ages. Nature 405(6789):907–913
11. Hewitt GM (2004) Genetic consequences of climatic oscillations in the quaternary. Philos Trans R Soc Lond Ser B Biol Sci 359(1442):183–195
12. Bintanja R, Van De Wal RS, Oerlemans J (2005) Modelled atmospheric temperatures and global sea levels over the past million years. Nature 437(7055):125–128
13. Garg KM, Chattopadhyay B, Wilton PR, Prawiradilaga DM, Rheindt FE (2018) Pleistocene land bridges act as semipermeable agents of avian gene flow in Wallacea. Mol Phylogenet Evol 125:196–203
14. Chattopadhyay B, Garg KM, Gwee CY, Edwards SV, Rheindt FE (2017) Gene flow during glacial habitat shifts facilitates character displacement in a Neotropical flycatcher radiation. BMC Evol Biol 17(1):1–15
15. Teeling EC, Vernes SC, Dávalos LM, Ray DA, Gilbert MTP, Myers E et al (2018) Bat biology, genomes, and the Bat1K project: to generate chromosome-level genomes for all living bat species. Annu Rev Anim Biosci 6:22–46

16. Chattopadhyay B, Garg KM, Vinoth KA, Ramakrishnan U, Kandula S (2012) Sibling species in South Indian populations of the rufous horse-shoe bat *Rhinolophus rouxii*. Conserv Genet 13(6):1435–1445

17. Mayer F, O H (2001) Cryptic diversity in European bats. Proc R Soc B 268(1478):1825–1832

18. Jones G, Van Parijs SM (1993) Bimodal echolocation in pipistrelle bats: are cryptic species present? Proc R Soc B 251(1331):119–125

19. Thabah A, Rossiter SJ, Kingston T, Zhang S, Parsons S, Mya KM et al (2006) Genetic divergence and echolocation call frequency in cryptic species of *Hipposideros larvatus* sl.(Chiroptera: Hipposideridae) from the Indo-Malayan region. Biol J Linn Soc 88(1):119–130

20. Jones G, Jacobs DS, Kunz TH, Willig MR, Racey PA (2009) Carpe noctem: the importance of bats as bioindicators. Endanger Species Res 8(1–2):93–115

21. Kunz TH, Fenton MB (eds) (2005) Bat ecology. University of Chicago Press, USA

22. Russo D, Ancillotto L (2015) Sensitivity of bats to urbanization: a review. Mamm Biol 80(3):205–212

23. O'Shea TJ, Cryan PM, Hayman DT, Plowright RK, Streicker DG (2016) Multiple mortality events in bats: a global review. Mammal Rev 46(3):175–190

24. Pruvot M, Cappelle J, Furey N, Hul V, Heng HS, Duong V et al (2019) Extreme temperature event and mass mortality of insectivorous bats. Eur J Wildl Res 65(3):1–5

25. Welbergen JA, Klose SM, Markus N, Eby P (2008) Climate change and the effects of temperature extremes on Australian flying-foxes. Proc R Soc B 275(1633):419–425

26. McCracken GF, Bernard RF, Gamba-Rios M, Wolfe R, Krauel JJ, Jones DN, Russell AL, Brown VA (2018) Rapid range expansion of the Brazilian free-tailed bat in the southeastern United States, 2008–2016. J Mammal 99(2):312–320

27. Kim S, Cho YS, Kim H-M, Chung O, Kim H, Jho S et al (2016) Comparison of carnivore, omnivore, and herbivore mammalian genomes with a new leopard assembly. Genome Biol 17(1):1–12

28. Chattopadhyay B, Garg KM, Mendenhall IH, Rheindt FE (2019) Historic DNA reveals Anthropocene threat to a tropical urban fruit bat. Curr Biol 29(24):R1299–R1300

29. Ramakrishnan U, Hadly EA, Mountain JL (2005) Detecting past population bottlenecks using temporal genetic data. Mol Ecol 14(10):2915–2022

30. Díez-del-Molino D, Sánchez-Barreiro F, Barnes I, Gilbert MTP, Dalén L (2018) Quantifying temporal genomic erosion in endangered species. Trends Ecol Evol 33(3):176–185

31. Mondol S, Bruford MW, Ramakrishnan U (2013) Demographic loss, genetic structure and the conservation implications for Indian tigers. Proc R Soc B 280(1762):20130496

32. Fages A, Hanghøj K, Khan N, Gaunitz C, Seguin-Orlando A, Leonardi M et al (2019) Tracking five millennia of horse management with extensive ancient genome time series. Cell 177(6):1419–1435

33. Bi K, Linderoth T, Vanderpool D, Good JM, Nielsen R, Moritz C (2013) Unlocking the vault: next-generation museum population genomics. Mol Ecol 22(24):6018–6032

34. Rogers RL, Slatkin M (2017) Excess of genomic defects in a woolly mammoth on Wrangel island. PLoS Genet 13(3):e1006601

35. Campbell P, Schneider CJ, Adnan AM, Zubaid A, Kunz TH (2004) Phylogeny and phylogeography of Old World fruit bats in the *Cynopterus brachyotis* complex. Mol Phylogenet Evol 33(3):764–781

36. Campbell P, Schneider CJ, Adnan AM, Zubaid A, Kunz TH (2006) Comparative population structure of *Cynopterus* fruit bats in peninsular Malaysia and southern Thailand. Mol Ecol 15(1):29–47

37. Bird MI, Taylor D, Hunt C (2005) Palaeoenvironments of insular Southeast Asia during the last glacial period: a savanna corridor in Sundaland? Quat Sci Rev 24(20–21):2228–2242

38. Bird MI, Pang WC, Lambeck K (2006) The age and origin of the straits of Singapore. Palaeogeogr Palaeoclimatol Palaeoecol 241(3–4):531–538

39. Brook BW, Sodhi NS, Ng PK (2003) Catastrophic extinctions follow deforestation in Singapore. Nature 424(6947):420–423

40. Carpenter ML, Buenrostro JD, Valdiosera C, Schroeder H, Allentoft ME, Sikora M et al (2013) Pulling out the 1%: whole-genome capture for the targeted enrichment of ancient DNA sequencing libraries. Am J Hum Genet 93(5):852–864

41. Jónsson H, Ginolhac A, Schubert M, Johnson PL, Orlando L (2013) mapDamage2.0: fast approximate Bayesian estimates of ancient DNA damage parameters. Bioinformatics 29(13):1682–1684

42. Johnson MG, Gardner EM, Liu Y, Medina R, Goffinet B, Shaw AJ et al (2016) HybPiper: extracting coding sequence and introns for phylogenetics from high-throughput sequencing reads using target enrichment. Appl Plant Sci 4(7):1600016

43. Korneliussen TS, Albrechtsen A, Nielsen R (2014) ANGSD: analysis of next generation sequencing data. BMC Bioinf 15(1):356

44. Borowiec ML (2016) AMAS: a fast tool for alignment manipulation and computing of summary statistics. PeerJ 4:e1660

45. Coulon A (2010) GENHET: an easy-to-use R function to estimate individual heterozygosity. Mol Ecol Resour 10(1):167–169

46. Team RC (2018) R: A language and environment for statistical computing

47. Wang J (2011) COANCESTRY: a program for simulating, estimating and analysing relatedness and inbreeding coefficients. Mol Ecol Resour 11(1):141–145

48. Purcell S, Neale B, Todd-Brown K, Thomas L, Ferreira MA, Bender D et al (2007) PLINK: a tool set for whole-genome association and population-based linkage analyses. Am J Hum Genet 81(3):559–575

49. Sodhi NS, Posa MRC, Lee TM, Bickford D, Koh LP, Brook BW (2010) The state and conservation of southeast Asian biodiversity. Biodivers Conserv 19(2):317–328

50. Sodhi NS, Koh LP, Clements R, Wanger TC, Hill JK, Hamer KC et al (2010) Conserving Southeast Asian forest biodiversity in human-modified landscapes. Biol Conserv 143(10):2375–2384

51. Wilcove DS, Giam X, Edwards DP, Fisher B, Koh LP (2013) Navjot's nightmare revisited: logging, agriculture, and biodiversity in Southeast Asia. Trends Ecol Evol 28(9):531–540

52. Myers N, Mittermeier RA, Mittermeier CG, Da Fonseca GA, Kent J (2000) Biodiversity hotspots for conservation priorities. Nature 403(6772):853–858

53. Corlett RT (1992) The ecological transformation of Singapore, 1819–1990. J Biogeogr 19:411–420

54. Castelletta M, Sodhi NS, Subaraj R (2000) Heavy extinctions of forest avifauna in Singapore: lessons for biodiversity conservation in Southeast Asia. Conserv Biol 14(6):1870–1880

55. Ming LT, Wai CK (2011) Bats in Singapore – ecological roles and conservation needs. In NSS symposium 2011: nature conservation for a sustainable Singapore, pp 41–64

56. Frankham R, Briscoe DA, Ballou JD (2002) Introduction to conservation genetics. Cambridge University Press, Cambridge

57. Saccheri I, Kuussaari M, Kankare M, Vikman P, Fortelius W, Hanski I (1998) Inbreeding and extinction in a butterfly metapopulation. Nature 392(6675):491–494

58. Feng S, Fang Q, Barnett R, Li C, Han S, Kuhlwilm M et al (2019) The genomic footprints of the fall and recovery of the crested ibis. Curr Biol 29(2):340–349

59. Hoelzel AR, Halley J, O'Brien SJ, Campagna C, Arnbom T, Le Boeuf B, Ralls K, Dover GA (1993) Elephant seal genetic variation and the use of simulation models to investigate historical population bottlenecks. J Hered 84(6):443–449

60. Hoelzel AR, Fleischer RC, Campagna C, Le Boeuf BJ, Alvord G (2002) Impact of a population bottleneck on symmetry and genetic diversity in the northern elephant seal. J Evol Biol 15(4):567–575

61. Excoffier L, Dupanloup I, Huerta-Sánchez E, Sousa VC, Foll M (2013) Robust demographic inference from genomic and SNP data. PLoS Genet 9(10):e1003905

62. Briskie JV, Mackintosh M (2004) Hatching failure increases with severity of population bottlenecks in birds. Proc Natl Acad Sci U S A 101(2):558–561

63. Chattopadhyay B (2018) Tales of the night: chapter I. CEiBa Newsl 1(3):14–19

Chapter 3
Are Molossid Bats Behaviourally Preadapted to Urban Environments? Insights from Foraging, Echolocation, Social, and Roosting Behaviour

Rafael Avila-Flores, Rafael León-Madrazo, Lucio Perez-Perez, Aberlay Aguilar-Rodríguez, Yaksi Yameli Campuzano-Romero, and Alba Zulema Rodas-Martínez

Abstract For most wildlife, cities are fairly hostile and resource-limited environments. However, some species of bats can persist and even thrive in cities. Species of Molossidae dominate urban bat assemblages in many tropical and subtropical cities around the world, but which intrinsic traits explain their ability to persist in cities? We explore how molossids may be preadapted to cities, with particular attention to their foraging, echolocation, social, and roosting behaviours. We hypothesise that behavioural plasticity strongly drives their ability to exploit the urban environment.

Keywords Preadaptation · Molossidae · Behavioural plasticity · Urban exploiters · Fast-flying bats

1 Introduction

Urbanisation is a human-driven process of land cover change that reduces the size and connectivity of natural and agricultural habitats and that modifies the biophysical attributes of the remaining, inner fragments. At finer scales, urban spaces represent novel ecosystems in which abiotic conditions, biotic interactions, and distribution and abundance of resources are all altered, to varying degrees, relative

R. Avila-Flores (✉) · R. León-Madrazo · L. Perez-Perez · A. Aguilar-Rodríguez · Y. Y. Campuzano-Romero · A. Z. Rodas-Martínez
División Académica de Ciencias Biológicas, Universidad Juárez Autónoma de Tabasco, Villahermosa, Tabasco, Mexico

© The Author(s), under exclusive license to Springer Nature Switzerland AG 2022 33
L. Moretto et al. (eds.), *Urban Bats*, Fascinating Life Sciences,
https://doi.org/10.1007/978-3-031-13173-8_3

to the native ecosystem. During the process of urbanisation, living organisms face novel environmental conditions that may represent a challenge or an opportunity depending on their specific traits. Within a population, natural selection favours those individuals whose morphological, physiological, and behavioural traits are better suited to face the novel, urban conditions. Consequently, evolutionary adaptation may gradually take place after many generations in animal populations living in cities [1].

Evolutionary adaptation, a process that occurs over many generations, is not the only mechanism that allows populations to successfully exploit urban environments. The phenotypes of organisms facing novel urban conditions may help to predict individual-, population-, and species-specific responses to urbanisation. Urban adaptedness, or the ability to immediately respond to challenges imposed by urban environments [1], is largely determined by pre-existing adaptations or preadaptations (also known as exaptations) of species. Among bats, insectivorous species with long, narrow, and pointed wings seem preadapted to successfully exploit urban areas [2, 3]. To explain this pattern, it has been hypothesised that a fast, economical flight allows individuals to readily move among patches of food within the city while also favouring the efficient exploitation of insects that swarm at streetlamps. In addition, fast-flying bats might avoid many dangers of the urban environment by flying at high altitudes [4, 5].

The family Molossidae (free-tailed bats), the fourth most diverse among Chiropteran families with 122 species [6], includes the most prominent examples of bats having relatively long, narrow, and pointed wings [7]. As expected, molossid species are among the most successful urban exploiters in the world [3], with relatively abundant urban populations in Africa (e.g. large-eared free-tailed bat, *Otomops martiensseni*), Europe (e.g. European free-tailed bat, *Tadarida teniotis*), Oceania (e.g. white-striped free-tailed bat, *Tadarida australis*), and the Americas (e.g. Mexican free-tailed bat, *Tadarida brasiliensis* and *Molossus* spp.). However, molossids' responses to urbanisation are species-specific and geographically variable. Not all molossids can be considered urban exploiters, and, indeed, some appear to be urban avoiders (e.g. hairless bat, *Cheiromeles torquatus*, and dwarf dog-faced bat, *Molossops temminckii*) [2, 8]. This begs the question of whether molossid bats are really preadapted to urban environments. What factors besides flight style may help to explain this family's differential responses to urbanisation? In this chapter, we explore the potential roles of foraging, echolocation, roosting, and social behaviours in molossid preadaptation to urbanisation and discuss the importance of behavioural plasticity to urban tolerance.

2 Foraging Behaviour

More than any other mammalian order, bats have extremely diverse morphofunctional adaptations that allow them to successfully forage in specific environmental contexts [9]. For example, insectivorous bats that have gracile skulls, thin mandibles, and small teeth tend to feed upon soft-bodied insects such as lepidopterans, whereas species that have crested skulls, thick mandibles, and larger teeth usually hunt hard-shelled insects, such as beetles [10]. Meanwhile, wing morphology largely determines which microhabitat (e.g. cluttered vs uncluttered, open vs edge) bats will use to forage [7]. Because the distribution of preferred food resources varies over space and time (even at small spatiotemporal scales), species' morphological traits may relate to other aspects of foraging behaviour, such as flight altitude [11], maximal distance travelled to forage [7], or temporal patterns of activity [which might be coupled with temporal availability of food; 12].

Molossids are the most remarkably adapted bats when it comes to hawking on high-flying insects. They have high aspect ratio (long and narrow) wings with high wing loading (heavy-bodied bats). This design may not confer the manoeuvrability needed in cluttered microhabitats, but it does promote fast and economical (low-energy expenditure) flight, which is well-suited to hunting in open spaces [7]. This flight style allows some species to move unusually long distances (≤ 50 km) at high altitude (> 3000 m) to reach suitable patches of food each night [13]. Molossids also exhibit high foraging efficiency (high capture success), which may explain why some species have extremely short periods of activity [14]. Furthermore, the tendency of many species to emerge from their roost around sunset/sunrise [12] suggests that high ambient light does not significantly impede their foraging and commuting. Support for this hypothesis comes from observations that moonlight may not affect foraging by some molossids [e.g. velvety free-tailed bats (*Molossus molossus*); 15].

Molossid bats could take advantage of some behavioural and morphofunctional traits to successfully forage in highly urbanised landscapes. For example, flying fast and economically could help them access distant patches of food across spatially large cities [4, 5, 16] or in non-urban (e.g. rural, natural) areas outside cities [17]. Furthermore, high-altitude flight might mitigate (or entirely eliminate) many urban challenges for bats, including human activity, vehicles, and noise. Light tolerance might partly explain why some molossid species thrive in heavily urbanised environments [17]. For example, the bimodal pattern of emergence (sunset/sunrise) observed in black mastiff bats (*Molossus nigricans*, previously known as *Molossus rufus* [18]) in natural habitats [15] has also been observed in urban settings – it likely helps this bat hunt crepuscular urban beetles [14]. Being able to tolerate bright light must also surely be a prerequisite to exploiting clusters of insects at streetlamps. Even species that do not emerge from their roosts until well after dark (e.g. big free-tailed bats, *Nyctinomops macrotis*) may forage in the open at streetlights in highly urbanised sites [4].

The above-mentioned traits (wing morphology and light tolerance) typify molossids, and although at first glance this might seem to make all molossid species preadapted to forage in cities, that is not the case. In fact, some species of generally urban-tolerant genera (*Tadarida, Otomops, Eumops, Molossus*) are rare in urban environments. As such, factors besides flight style must explain urban tolerance among molossid bats. Future research could analyse, for example, whether and how dietary plasticity helps bats to hunt novel urban insect assemblages, the importance of vertical and horizontal distribution of potential prey in cities, or the role of learning in detecting high-quality food patches that are ephemeral and unpredictable (e.g. insect cluster swarms at stadium lamps). Although these lines of inquiry are worthwhile, they are challenging to pursue because of how difficult it is to record bats at high altitudes.

3 Echolocation Behaviour

Echolocation is used by most bats to create a three-dimensional image of the surrounding environment and involves the emission of high-frequency vocalisations (calls) and the subsequent decoding of the returning echoes [19]. Bats use echolocation to deal with the challenges of nocturnality and specifically to detect, track, and capture prey (or forage on other foods), to navigate and avoid obstacles, and to create acoustic landmarks. Echolocation, in combination with social calls, can be used to transfer information (e.g. about sources of food) between group members inside and outside roosts and plays a role in mother-pup communication and mate selection [20]. Call design is not a fixed attribute, and it may be adjusted by echolocating bats depending on the immediate task and spatial context [21].

Molossids (but not *Molossops* spp.) typically emit long duration, low-frequency (20–40 kHz), search-phase echolocation calls with narrow bandwidth and long interpulse intervals – a strategy that lets them detect large insects in the open over savannas and forest canopies [7, 22], but not discriminate objects against cluttered backgrounds [23]. However, individual *T. brasiliensis* have been shown to greatly modify call design during mass emergence from cave roosts, increasing both the bandwidth and the frequency to get a more detailed picture of their surroundings [24]. Other molossids also exhibit a high call plasticity, especially when it comes to adjusting the shape, frequency, and bandwidth, which might represent an "advanced evolutionary trait" of these bats, as argued by Jung, Molinari, and Kalko [22].

In combination with wing morphology, echolocation call design is identified as a key predictor of bats' urban tolerance. Urban tolerance is correlated with lower (peak or characteristic) call frequencies among Asian, Australian, and North American bats and longer call durations among Australian and South American bats [2]. The ability to detect distant objects, coupled with a fast and economical flight, allows molossid bats to move and hunt in the open above buildings, noisy areas, and other urban features. Compared to slow-flying insectivores, molossids might rely less on acoustic landmarks on the ground to orientate and commute in cities [4].

The fact that not all molossids succeed in cities raises the question of whether echolocation behaviour differs between urban-tolerant and urban-sensitive species. Species of *Molossops*, for example, would not be expected to succeed in highly urbanised sites because they produce higher-frequency, shorter calls than other molossids [22]. Following the conclusions of Jung and Threlfall [2], it is likely that larger molossids that emit lower-frequency calls are better preadapted to forage in open spaces above urban features than smaller species. However, some ability to detect trees, power cables, poles, and even other foraging bats at relatively short distances would help molossid bats forage in the proximity of streetlights. Such an ability would also be useful to detect and avoid obstacles when bats emerge from or return to roosts, especially roosts in low-rise structures. In addition, the capacity to adjust call frequency would reduce interference from urban noise. Therefore, we hypothesise that vocal plasticity is an important determinant of a species' preadaptation to urban landscapes. In fact, high variation in the echolocation call structure of *T. brasiliensis* (an urban-tolerant bat) has been reported for individuals flying at different altitudes, from a few to more than 800 m above the ground [25]. Adjustments in call frequencies have also been reported for individual *T. brasiliensis* foraging in the presence of conspecifics at urban streetlights [26]. However, the study of vocal plasticity in urban bats is a research area that has received little attention.

4 Social Behaviour

Most species of bats spend most of their lives in groups, exhibiting a remarkable diversity of social organisation systems (including mating, roosting, and foraging systems). Although most bat groupings gather exclusively inside roosts, they may also do so outside roosts for reproductive or foraging purposes. Within roosts, bats exhibit diverse group forms that can be characterised based on their structure, composition, and sociality. According to Kerth [27], groupings may involve large numbers of individuals that are rarely in physical contact (aggregations), moderate numbers of individuals that are usually in physical contact (colonies), or smaller numbers of bats which direct more affiliative or sexual interactions towards their own group mates than towards members of neighbour groups (social groups). The benefits of group life for individual bats include thermoregulatory efficiency (and, hence, energy savings), optimisation of pup growth, information exchange, enhanced detection of suitable roosts, defence of mating partners, and, in some cases, enhanced cooperative behaviours such as allogrooming and food exchange [28].

Molossids exhibit a variety of grouping strategies within day roosts. In natural environments, group sizes may range from tens [29] to millions [30] of individuals, depending on roost permanency, roost availability, and the species' social system. Within roosts, some molossids, such as Florida bonneted bats (*Eumops floridanus*), giant mastiff bat (*O. martiensseni*), Midas free-tailed bats (*Tadarida midas*), and little free-tailed bats (*Trichopsis pumila*), exhibit polygyny [31, 32], while others, such as broad-eared bats (*Nyctinomops laticaudatus*) and *T. brasiliensis*, are promiscuous [31, 33]. On the other hand, the nature and intensity of molossids' social interactions are poorly studied [except for the well-documented mother-pup bonding *T. brasiliensis*; 34]. In many molossids, chemical signals produced by gular glands (especially in males) seem to play a prominent role in the selection of sexual partners and/or the marking of territory during the mating season [35, 36]. Interestingly, studies on *E. floridanus* and *T. brasiliensis* suggest that the role of gular gland secretions of molossids may change depending on the size of the group-ing to optimise energy expenditure associated with resource defence [32, 37].

Cities potentially offer molossids many available roosts (albeit smaller ones compared to natural settings), in which it is possible to establish social groups or colonies of varying sizes [37]. Groups of *T. brasiliensis*, *M. nigricans*, Sinaloan mastiff bats (*Molossus sinaloae*), and dwarf bonneted bats (*Eumops bonariensis*) are relatively common in Neotropical cities [30, 35, 38], where they exhibit colony or group sizes like those in natural settings (except for *T. brasiliensis*, whose urban colonies tend to be much smaller than cave colonies). Scattered information in the literature suggests that the social structure of urban molossids is relatively flexible, with either male- or female-biased sex ratios [e.g. *Eumops glaucinus*, *M. nigricans*, *E. bonariensis*, *M. sinaloae*, *N. laticaudatus*, *T. teniotis*, and *T. brasiliensis* and big crested mastiff bat, *Promops centralis*; 35, 38]. In *T. brasiliensis*, social flexibility might extend to the mating system, as small groups kept in captivity may exhibit polygyny with territorial defence [39] – an observation which suggests that some urban molossids might adjust their social structure to group size (which is in turn limited by roost dimensions). Furthermore, the fact that many molossids have gular glands speaks to the potential for males to defend territories in small roosts like those commonly encountered in cities. We suggest that molossids' flexible social organisation (as a function of group size) could be an additional factor that favours some species in urban environments. However, the limited information on the social system of most molossid species [31] makes this hypothesis difficult to test.

5   Roosting Behaviour

Because bats spend most of their lives roosting, the selective pressures of the roost environment are of paramount importance. A suitable roost offers protection against predators and the elements, favours social interactions, facilitates the care and growth of pups, and reduces the energetic costs of thermoregulation [40]. It has been suggested that microclimate (temperature, humidity, and wind speed) is the most important determinant of roost selection in bats [41], and for many species, roost permanency affects roost fidelity [42]. Therefore, it is not surprising that caves (permanent structures with stable microclimates) are a common type of roost for most bats, although many species commonly use permanent anthropogenic struc-tures, such as mines, buildings, roofs, culverts, and bridges.

Except for some species (e.g. *T. brasiliensis*), the natural roosting behaviour of most molossids is poorly known. There is scattered evidence that molossids in natu-ral environments use diverse roosts, including caves [43], tree cavities [44], crevices in rocky cliffs [45], or even holes in archaeological ruins [33]. Most molossids do not regularly roost in caves, but when they do, they can form huge aggregations (e.g. *T. brasiliensis* and *N. laticaudatus*). Depending on the season, *T. brasiliensis* may use caves as mating roosts, maternity roosts, or unisexual non-reproductive roosts [46]. In this review, we did not find any studies of roost selection (contrasting used with unused/available roosts) by molossids in natural areas, which likely reflects the natural scarcity of suitable roosts for these bats. It is also likely that most molossid roosts are in high, open areas (e.g. crevices in cliffs or holes in tall trees) that are hard for researchers to find and access. High entrances with open surroundings could be an obvious requirement for fast-flying bats that must drop down into flight from the roost entrance [14]. Therefore, it is not surprising that anthropogenic structures near natural areas are heavily used by molossid bats even when some natural roosts may be available [47].

Some morphological traits may preadapt molossids to exploit anthropogenic roosts in urban and non-urban environments. Some species are dorsoventrally flattened, which facilitates occupancy of narrow crevices, including inbuilt structures [48, 49]. The posterior flexion of the first phalanxes of wing digits three and four, in combination with robust and muscular forelimbs, helps them crawl rapidly on the ground, on walls, and inside narrow spaces. In addition, the thick patagium could make molossid wings more resistant to constant friction with coarse or rocky surfaces in narrow spaces [50].

The commonality of the above-mentioned morphological traits among all molossids might seem to suggest that they are all preadapted to occupy narrow crevices with high entrances in urban buildings. However, the ability to occupy urban roosts seems to vary among species. *T. brasiliensis*, probably the most abundant urban bat in the New World, prefers abandoned buildings that have large and high roost entrances and are surrounded by scarce vegetation [37]. In contrast, western mastiff bats (*Eumops perotis*), the largest molossids, usually occupy attics and may take off from exceptionally low heights (≥ 2 m). Therefore, the height of the roost entrance may not be as limiting for molossids as previously thought. Plasticity, when it comes to roost microclimate requirements, may offer another clue to the urban success of some species. For example, Angolan free-tailed bats (*Mops condylurus*) roost in narrow spaces under metal sheet roofs in small houses (spaces that often exceed 40 °C) and yet can enter torpor when ambient temperature falls to 15 °C; this unique plasticity makes *M. condylurus* well suited to occupy a variety of man-made roosts [51]. A similar tolerance of warm roosts has been reported for *T. brasiliensis* [51]. Another important preadaptation that maximises the roosting plasticity of some molossids (e.g. *M. molossus*, *M. nigricans*, *T. teniotis*, and little northern free-tailed bats, *Mormopterus loriae*) is the ability to enter torpor-like physiological states at moderate and relatively high ambient temperatures [51, 52]. No doubt, detailed studies of roost selection in natural and urban contexts will elucidate which factors promote the urban success of some molossid bats.

6 Conclusions

Global meta-analyses using ecological, behavioural, and life history traits have identified increased mobility as the most important factor explaining the success of molossid bats in urban landscapes. Here, we suggest that some foraging, echolocation, social, and roosting behavioural traits contribute to preadapt many molossid species to persist in highly urbanised environments. These traits include (1) high-altitude flight, (2) light tolerance, (3) lack of reliance on acoustic landmarks for orientation, (4) vocal plasticity (to detect near and distant objects), (5) the ability to adjust social systems to group size, (6) the ability to occupy narrow spaces in diverse built elements, and (7) physiological tolerance to a wide range of roost temperatures. Combined, these traits may make some species preadapted to urbanisation. Because so much information on the behavioural ecology of molossid bats is anecdotical and still scarce, our conclusions should be interpreted as hypotheses that need to be tested.

Literature Cited

1. McDonnell MJ, Hahs AK (2015) Adaptation and adaptedness of organisms to urban environments. Annu Rev Ecol Evol Syst 46(1):261–280
2. Jung K, Threlfall CG (2018) Trait-dependent tolerance of bats to urbanization: a global meta-analysis. Proc R Soc B Biol Sci 285(1885):20181222
3. Santini L, González-Suárez M, Russo D, Gonzalez-Voyer A, von Hardenberg A, Ancillotto L (2019) One strategy does not fit all: determinants of urban adaptation in mammals. Ecol Lett 22(2):365–376
4. Avila-Flores R, Fenton MB (2005) Use of spatial features by foraging insectivorous bats in a large urban landscape. J Mammal 86(6):1193–1204
5. Jung K, Kalko EKV (2011) Adaptability and vulnerability of high flying Neotropical aerial insectivorous bats to urbanization. Divers Distrib 17(2):262–274
6. Simmons NB (2005) Order Chiroptera. In: Mammal species of the world: a taxonomic and geographic reference, 3rd edn. Johns Hopkins University Press, pp 312–529
7. Norberg UM, Rayner JMV, Lighthill MJ (1987) Ecological morphology and flight in bats (Mammalia; Chiroptera): wing adaptations, flight performance, foraging strategy and echolocation. Philos Trans R Soc London B, Biol Sci 316(1179):335–427
8. Jung K, Threlfall CG (2016) Urbanisation and its effects on bats – a global meta-analysis. In: Voigt CC, Kingston T (eds) Bats in the anthropocene: conservation of bats in a changing world. Springer International Publishing, Cham, pp 13–33
9. Denzinger A, Schnitzler H-U. Bat guilds, a concept to classify the highly diverse foraging and echolocation behaviors of microchiropteran bats. Front Physiol 2013;4, 164
10. Freeman PW (1981) Correspondence of food habits and morphology in insectivorous bats. J Mammal 62(1):166–173
11. Roemer C, Coulon A, Disca T, Bas Y (2019) Bat sonar and wing morphology predict species vertical niche. J Acoust Soc Am 145(5):3242–3251
12. Jones G, Rydell J (1994) Foraging strategy and predation risk as factors influencing emergence time in echolocating bats. Philos Trans R Soc London Ser B Biol Sci 346(1318):445–455
13. McCracken GF, Lee Y-F, Gillam EH, Frick W, Krauel J (2021) Bats flying at high altitudes. In: Lim BK, Fenton MB, Brigham RM, Mistry S, Kurta A, Gillam EH et al (eds) 50 years of bat research. Springer International Publishing, Cham, pp 189–205
14. Fenton MB, Rautenbach IL, Rydell J, Arita HT, Ortega J, Bouchard S et al (1998) Emergence, echolocation, diet and foraging behavior of *Molossus ater* (Chiroptera: Molossidae). Biotropica 30(2):314–320
15. Holland RA, Meyer CFJ, Kalko EKV, Kays R, Wikelski M (2011) Emergence time and foraging activity in Pallas' mastiff bat, *Molossus molossus* (Chiroptera: Molossidae) in relation to sunset/sunrise and phase of the moon. Acta Chiropterologica 13(2):399–404

16. Rhodes M, Catterall C (2008) Spatial foraging behavior and use of an urban landscape by a fast-flying bat, the Molossid *Tadarida australis*. J Mammal 89(1):34–42

17. Jung K, Kalko EKV (2010) Where forest meets urbanization: foraging plasticity of aerial insectivorous bats in an anthropogenically altered environment. J Mammal 91(1):144–153

18. Loureiro LO, Engstrom MD, Lim BK (2020) Single nucleotide polymorphisms (SNPs) provide unprecedented resolution of species boundaries, phylogenetic relationships, and genetic diversity in the mastiff bats (*Molossus*). Mol Phylogenet Evol 143:106690

19. Neuweiler G (2000) The biology of bats. Oxford University Press

20. Pfalzer G, Kusch J (2003) Structure and variability of bat social calls: implications for specificity and individual recognition. J Zool 261(1):21–33

21. Obrist MK (1995) Flexible bat echolocation: the influence of individual, habitat and conspecifics on sonar signal design. Behav Ecol Sociobiol 36(3):207–219

22. Jung K, Molinari J, Kalko EKV (2014) Driving factors for the evolution of species-specific echolocation call Design in new World Free-Tailed Bats (Molossidae). PLoS One 9(1):e85279

23. Schnitzler H-U, Kalko EKV (2001) Echolocation by insect-eating bats: we define four distinct functional groups of bats and find differences in signal structure that correlate with the typical echolocation tasks faced by each group. Bioscience 51(7):557–569

24. Gillam EH, Hristov NI, Kunz TH, McCracken GF (2010) Echolocation behavior of Brazilian free-tailed bats during dense emergence flights. J Mammal 91(4):967–975

25. Gillam EH, McCracken GF, Westbrook JK, Lee Y-F, Jensen ML, Balsley BB (2009) Bats aloft: variability in echolocation call structure at high altitudes. Behav Ecol Sociobiol 64(1):69–79

26. Ratcliffe JM, HMT H, Avila-Flores R, Fenton MB, McCracken GF, Biscardi S et al (2004) Conspecifics influence call design in the Brazilian free-tailed bat, *Tadarida brasiliensis*. Can J Zool 82(6):966–971

27. Kerth G (2008) Causes and consequences of sociality in bats. Bioscience 58(8):737–746

28. Wilkinson GS, Carter G, Bohn KM, Caspers B, Chaverri G, Farine D et al (2019) Kinship, association, and social complexity in bats. Behav Ecol Sociobiol 73(1):1–15

29. Ober HK, Braun De Torrez EC, Gore JA, Bailey AM, Myers JK, Smith KN et al (2017) Social organization of an endangered subtropical species, *Eumops floridanus*, the Florida bonneted bat. Mammalia 81(4):375–383

30. Weaver SP, Simpson TR, Baccus JT, Weckerly FW (2015) Baseline population estimates and microclimate data for newly established overwintering Brazilian free-tailed bat colonies in Central Texas. Southwest Nat 60(2–3):151–157

31. McCracken GF, Wilkinson GS (2000) Bat mating systems. In: Crichton EG, Krutzsch PH (eds) Biology of bats. Academic Press, London, pp 321–362

32. Braun de Torrez EC, Gore JA, Ober HK (2020) Evidence of resource-defense polygyny in an endangered subtropical bat, Eumops floridanus. Glob Ecol Conserv 24:e01289

33. Ortega J, Hernández-Chávez B, Rizo-aguilar A, Guerrero JA (2010) Social structure and temporal composition in a colony of *Nyctinomops laticaudatus*. Rev Mex Biodivers 81(3):853–862

34. McCracken GF, Gustin MK (1991) Nursing behavior in Mexican free-tailed bat maternity colonies. Ethology 89(4):305–321

35. Keeley ATH, Keeley BW (2004) The mating system of *Tadarida brasiliensis* (Chiroptera: Molossidae) in a large highway bridge colony. J Mammal 85(1):113–119

36. Englert AC, Greene MJ (2009) Chemically-mediated roostmate recognition and roost selection by Brazilian free-tailed bats (*Tadarida brasiliensis*). PLoS One 4(11):e7781

37. Li H, Wilkins KT (2015) Selection of building roosts by Mexican free-tailed bats (*Tadarida brasiliensis*) in an urban area. Acta Chiropterologica 17(2):321–330

38. Bowles JB, Heideman PD, Erickson KR (1990) Observations on six species of free-tailed bats (Molossidae) from Yucatan, Mexico. Southwest Nat 35(2):151

39. French B, Lollar A (1998) Observations on the reproductive behavior of captive *Tadarida brasiliensis mexicana* (Chiroptera: Molossidae). Southwest Nat 43(4):484–490

40. Kunz TH (1982) Roosting ecology of bats. In: Kunz TH (ed) Ecology of bats. Springer, Boston, pp 1–55

41. Ávila-Flores R, Medellín RA (2004) Ecological, taxonomic, and physiological correlates of cave use by Mexican bats. J Mammal 85(4):675–687
42. Lewis SE (1995) Roost fidelity of bats: a review. J Mammal 76(2):481–496
43. Allen LC, Turmelle AMYS, Widmaier EP, Hristov NI, McCracken GF, Kunz TH (2011) Variation in physiological stress between bridge- and cave-roosting Brazilian free-tailed bats. Conserv Biol 25(2):374–381
44. Rhodes M, Wardell-Johnson G (2006) Roost tree characteristics determine use by the white-striped freetail bat (*Tadarida australis,* Chiroptera: Molossidae) in suburban subtropical Brisbane, Australia. Austral Ecol 31(2):228–239
45. Corbett RJM, Chambers CL, Herder MJ (2008) Roosts and activity areas of *Nyctinomops macrotis* in northern Arizona. Acta Chiropterologica 10(2):323–329
46. Allen LC, Richardson CS, McCracken GF, Kunz TH (2010) Birth size and postnatal growth in cave- and bridge-roosting Brazilian free-tailed bats. J Zool 280(1):8–16
47. Randrianandrianina F, Andriafidison D, Kofoky AF, Ramilijaona O, Ratrimomanarivo F, Racey PA et al (2006) Habitat use and conservation of bats in rainforest and adjacent human-modified habitats in eastern Madagascar. Acta Chiropterologica 8(2):429–437
48. López-Baucells A, Rocha R, Andriatafika Z, Tojosoa T, Kemp J, Forbes K et al (2017) Roost selection by synanthropic bats in rural Madagascar: what makes non-traditional structures so tempting? Hystrix Ital J Mammal 28(1):28–35
49. Altringham JD (2011) Bats: from evolution to conservation, 2nd edn, Oxford, p 324
50. Vaughan TA (1966) Morphology and flight characteristics of Molossid bats. J Mammal 47(2):249–260
51. Licht P, Leitner P (1967) Behavioral responses to high temperatures in three species of California bats. J Mammal 48(1):52–61
52. Teague O'Mara M, Rikker S, Wikelski M, Ter MA, Pollock HS, Dechmann DKN (2017) Heart rate reveals torpor at high body temperatures in lowland tropical free-tailed bats. R Soc Open Sci 4(12)

Chapter 4
Urban Bats and their Parasites

Elizabeth M. Warburton, Erin Swerdfeger, and Joanna L. Coleman

Abstract Understanding host-parasite relationships in urban environments provides information critical for understanding bat ecology in anthropogenically altered landscapes. Although most current evidence comes from bat-virus systems, links between bats and their ectoparasites and endoparasites can provide key examples of how anthropogenic change affects bat health, roosting and foraging ecology, and, ultimately, bat conservation. This chapter examines the current state of knowledge and identifies potentially understudied aspects of urban bats and their parasites. Urbanisation can potentially modulate bat-parasite associations by affecting resource availability, ecophysiology, behaviour, and life history of bats. Urbanisation may also influence how these effects vary among parasites, bat species, and bat age classes. We distinguish between the effects of urbanisation in relation to ectoparasites and endoparasites, with one illustrative case study of each. The first case study examines ectoparasites of little brown bats (*Myotis lucifugus*) along an urban-rural gradient. It found some indications that *M. lucifugus* were more heavily parasitised in the city, likely because this was where the bats were most abundant and because ectoparasitism often rises along with host population density. The second case study investigates how anthropogenic habitat disturbance contributes to shifting helminth communities in big brown bats (*Eptesicus fuscus*). Land cover categories with more intense human activities were most likely to have similar helminth communities, likely because worms that parasitise more ecologically sensitive, intermediate hosts are more prone to extirpation with increasing anthropogenic disturbance. Finally, we conclude by suggesting that the tightly linked nature of the host-parasite relationship provides unique opportunities to

E. M. Warburton (✉)
Center for the Ecology of Infectious Diseases, Odum School of Ecology,
University of Georgia, Athens, GA, USA

E. Swerdfeger
Department of Biology, University of Regina, Regina, SK, Canada

J. L. Coleman
Department of Biology, Queens College at the City University of New York,
Queens, NY, USA

© The Author(s), under exclusive license to Springer Nature Switzerland AG 2022   43
L. Moretto et al. (eds.), *Urban Bats*, Fascinating Life Sciences,
https://doi.org/10.1007/978-3-031-13173-8_4

address key urban ecology questions related to host foraging and roosting in urban areas, host-vector contact rates in disturbed habitat, and host susceptibility in response to anthropogenic stressors.

Keywords Endoparasites · Ectoparasites · Vectors · Chiroptera · Urbanisation

1 Introduction

As cities expand and human populations in urban areas progressively outnumber those in rural areas, bats, like other wildlife, must increasingly contend with more frequent and intense human activities in urban areas relative to other land uses. These activities can drastically alter the amount, configuration, and quality of habitat for bats, especially via replacement of vegetation by built cover. As such, they can also alter bats' relationships with their parasites.

While early definitions of parasitism focus on trophic implications (i.e. parasites 'feed off' their hosts and often include the concept of harming their hosts [e.g. Crofton 1971, cited in [1]]), this trophic focus is a narrow view of parasitism. Indeed, parasites constitute not one but many taxa and must be at least somewhat adapted to their hosts (i.e. to evade immune responses [1]). Thus, parasitism may be viewed through the lens of hosts as habitat, with food sources located within the host habitat [1]. Some of these host-parasite relationships are visually dramatic when encountered in nature, such as with ectoparasites (Fig. 4.1), whereas endoparasites remain hidden inside their bat hosts. This chapter adopts this ecological-and evolutionary-based definition of parasitism.

Studying bats and their host-parasite relationships in cities can provide critical ecological information, such as selection of foraging and roosting sites, that directly impacts bat conservation, as urbanisation is a key extinction threat for bats [2]. Parasitism of wildlife generally is modulated by diverse environmental parameters and linked to the distribution, population dynamics, and health of hosts [3] – all of which may vary with urbanisation. For bats, much evidence comes from virus-related research. For example, certain flying foxes (*Pteropus* spp.) exhibit dramatic behavioural adjustments to recent land use and land cover change in Australia [4, 5]. In their ancestral forest habitat, they migrate over long distances searching for spatially and temporally patchy food resources. However, amid deforestation pushing them out of forests, they have been drawn into cities, where cultivated (native and exotic) trees offer fruit and/or nectar year-round. These conditions favour sedentary behaviour so that there are now permanent, large, aggregations of flying foxes in many cities, where none existed historically [4]. Additionally, pteropodids are the natural reservoirs for Hendra virus (HeV), and modelling suggests that these altered behaviours drive disease dynamics [6]. As bats become more urbanised and sedentary, connectivity between local populations and herd immunity across the metapopulation declines. This results in more sporadic but more intense outbreaks in

Fig. 4.1 This wingless bat fly (Nycteribiidae: *Penicillidia* sp.) nearly covers the entire face of a Mozambican long-fingered bat (*Miniopterus mossambicus*). (Photo credit: Dr. Piotr Naskrecki, Minden Pictures)

urban populations – a phenomenon with significant medical and veterinary implications.

Though the above example focuses on a virus, it illustrates the generalised, expected effect of higher host densities increasing host-parasite contact rates [7 and others therein], including parasites that present zoonotic disease risks. Pteropodids are not the only bats whose distributions and population dynamics may vary with urbanisation, which may ultimately influence bat-parasite dynamics. Other species may be more abundant in cities, especially synanthropes that readily exploit subsidised food resources or anthropogenic roosts [see also 8]. Moreover, species such as Kuhl's pipistrelle [*Pipistrellus kuhlii*; 9] and little brown bats [*Myotis lucifugus*; 10] may exhibit altered fecundity in relation to urbanisation – a phenomenon that could modulate bat-parasite dynamics by altering the relative availability of pups and juveniles. The naïve immune systems of young bats and their reduced ability to self-groom may make them more susceptible to parasitism. Alternatively, parasites may prefer adult hosts, given their higher overwintering survival [11 and others therein]. Thus, altered host age structure could affect parasite populations.

Urban-associated pollutants and other stressors in cities may compromise immune function or other aspects of host health with possible parasitological consequences. For example, adult female and juvenile *Pipistrellus kuhlii* foraging over a more polluted reservoir in the Negev Desert harboured more ectoparasites compared to individuals foraging over cleaner ponds [12]. However, the focus of this study was not urban pollutants, and a lack of site replication makes it difficult to conclusively attribute differential parasitism to water quality. Other taxa offer additional evidence of these phenomena. Serieys et al. [13] investigated the causes of a deadly outbreak of Notoedric mange, a parasitic skin disease that decimated an urban population of bobcats (*Felis rufus*) in California, United States of America (USA). Comparison of blood samples from *F. rufus* along an urbanisation gradient showed that exposure to rodenticides and urban land use was linked to reduced immune function and skin health and higher susceptibility to mange.

Although understanding bat-parasite relationships is relevant to bat biology and ecology, only 21% of 570 publications on bat parasites identified in our literature search were in the topic areas (defined by Web of Science) of ecology and conservation (Fig. 4.2). Other dominant research foci and/or motivations were biodiversity discovery and phylogeny (39%) and zoonoses (19%). Studies of these associations in urban areas are rare, i.e. a total of 29 potentially relevant papers – all but 4 published since 2015 and strongly biased towards the Neotropics (18 studies) and zoonotic questions (16 studies). The nearly universal approach has been to document parasites of bats in urban areas and sometimes compare findings with published data from non-urban areas, as opposed to conducting simultaneous comparisons (e.g. along urban gradients), which could help elucidate the urban ecology of bats and their parasites. As such, this chapter examines the current state of knowledge and identifies potentially understudied aspects of urban bats and their parasites. The central theme is urbanisation modulating the dynamics of bat-parasite associations via its effects on resource availability, ecophysiology, behaviour, and life history of bats. These impacts can vary among parasites, bat species, and bat cohorts; consequently, bat ectoparasites and endoparasites are discussed in detail with one illustrative case study each.

2   Ectoparasites

Bats host a huge diversity of ectoparasitic arthropods that spend their whole lives on the outside of bats' bodies and/or in their roosts and often have high host specificity [14]. Thus, their diversity and abundance are inexorably linked to aspects of bat

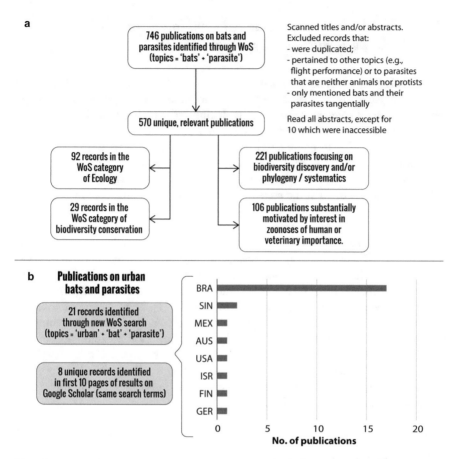

Fig. 4.2 Results of a literature search on 27 April 2021. (**a**) In step 1, we searched Web of Science (WoS) with the search terms 'bats' and 'parasite' in the topics. Scanning all titles and abstracts allowed us to exclude 176 records. We then read the 560 accessible abstracts of the resulting unique, relevant records and examined their distribution among WoS categories (in WoS analytics) and 2 research foci: (1) biodiversity discovery and phylogeny/systematics (i.e. species checklists, taxonomic revisions, evolution) and (2) zoonoses of human and veterinary importance, i.e. abstracts prominently mention diseases of humans, pets, or livestock). (**b**) In step 2, we performed a new WoS search with the search terms 'urban', 'bat', and 'parasite' in the topics and supplemented this with a Google Scholar search to identify other unique records (not indexed in WoS) in the first ten pages of results. We retained studies that reported original fieldwork (as opposed to meta-analyses) and classified these by location to explore geographic clustering of studies. (Histogram, country ISO codes on y-axis)

health, ecology, and behaviour that should be responsive to urbanisation. Additionally, various urban-related environmental changes (e.g. climate, pollution) may affect these ectoparasites independently of their hosts. Finally, these parasites may be disease vectors. Therefore, studying ectoparasitism in relation to urbanisation could help answer timely questions in urban ecology and bat roost selection, two key components of bat conservation.

2.1 What Are the Parasitological Consequences of Altered Roosting Behaviours By Urban Bats?

In most biomes, urban development reduces the availability of natural roosts while increasing that of anthropogenic structures. Therefore, any bat species' urban adaptedness is at least somewhat predicted by flexible roosting habits. Indeed, divergent behaviours between urban and non-urban bat populations are well-documented. These include shifts to commensal roosting, as in Brazil, where 84 species that inhabit cities mainly use built elements, especially buildings [15]. Other shifts include increased roost fidelity, as in the case of Indiana bats (*Myotis sodalis*), which switch less often at the edge of a developing urban area than in contiguous forest [16].

Both shifts have parasitological implications. First, not only does frequent roost switching correlate with reduced ectoparasitism, which may have evolved as an anti-parasitic strategy, but also this roost-switching declines in commensal roosts [7 and others therein]. Next, commensal roosts, being generally larger and more permanent than natural ones, favour larger colonies and tighter social networks among bats, and thus parasite transfers between individuals [7]. They also promote reproduction of insects (e.g. bat flies: Diptera, Streblidae, Nycteribiidae) that complete part of their life cycles in roosts [17]. Finally, ectoparasites may exhibit greater host specificity in commensal roosts occupied by a single bat species. For instance, four bat fly species parasitise a single species in Singapore, where their bat hosts use commensal roosts, but use multiple hosts elsewhere in Southeast Asia, where they roost with other bat species in caves [17].

2.2 Does Urbanisation Have Linked Fitness and Parasitological Implications for Bats?

Urbanisation could modulate either the prevalence or the intensity of ectoparasitism by impacting various indicators of bat fitness or affect bat fitness by modulating ectoparasitism factors. One indicator of this modulation is bat body condition, which may vary with urbanisation [3, 10] and often correlates with ectoparasitism – sometimes positively [e.g. various parasites on *M. lucifugus*; 18], sometimes negatively [e.g. bat flies on fruit bats; 17]. Yet, while the ectoparasites clearly gain resources, for example, by consuming the blood or lymph of their hosts, whether they directly and substantially reduce body condition is debatable [14] because establishing cause and effect is difficult. For example, finding that bats in better body condition harbour fewer parasites could indicate either that fitter individuals are better able to cope with parasites (e.g. have more energy to groom) or that they are not preferred hosts.

Another indicator of fitness that could vary with urbanisation is reproductive output. Evidence remains scant, but in Italy, as urban land cover around building maternity roosts of *Pipistrellus kuhlii* increases, so do numbers of pups per female [9]. Higher urban proportions of immature bats could, as mentioned, have either positive or negative effects on ectoparasitism levels, depending on their host age-class preferences. Additionally, these *P. kuhlii* give birth earlier in more urbanised roosts [9]. For temperate zone bats, earlier parturition is a fitness gain – it leaves more time for mothers and juveniles to accumulate fat reserves needed to overwinter. Earlier parturition could also be detrimental to various ectoparasites by reducing their optimal reproduction window. For example, two nycteribiid flies, one wing mite (*Spinturnix psi*) and one hard tick (*Ixodes simplex simplex*), on Schreiber's bat (*Miniopterus schreibersii*) in Portugal mainly reproduce on adult females and volant pups and mostly during pregnancy and lactation [19]. This is likely because pregnant females and pups have reduced behavioural and immune defences, and lactation enhances opportunities for vertical transmission while reducing the mother's available energy to groom [19].

2.3   Do Urban Abiotic Changes Modulate Bat-Ectoparasite Relationships?

Compared to surrounding areas, most cities are warmer and less humid, with altered precipitation and dampened seasonality – this is the urban heat island (UHI). For ectoparasites of bats, especially ones that live part of their lives off their hosts, such shifts could alter survival, reproduction, and/or host-seeking behaviour. Though this possibility has not been tested specifically in relation to the UHI, temperature and precipitation do affect bat flies parasitising bats in Venezuela, albeit differentially depending on the bat species [20]. Additionally, the UHI in Poland seemed linked to reduced abundance of *Ixodes ricinus* [21], which rarely parasitise bats but are in the same genus as other hard ticks that do.

Cities also tend to have high levels of various forms of pollution. One is heavy metal contamination, and evidence from a small sample of Daubenton's bats (*Myotis daubentonii*) in Finland [22] suggests that it might disrupt bat-ectoparasite associations. The likelihood that an individual harboured wing mites rose with its cadmium and copper exposures but declined with lead exposure, while arsenic and cobalt levels were negatively correlated with the presence of bat flies. Another urban issue is the presence of light and noise pollution. The implications for bats and their ectoparasites are unknown but may be worth studying given strong evidence that both stressors disrupt associations between túngara frogs (*Engystomops pustulosus*) and the *Corethrella* midges that bite them, namely, by reducing midge abundance [23].

2.4   Could Urban Changes in Ectoparasite Loads Alter the Risk of Disease?

Bats are a species-rich order [24] and, as such, host a wide diversity of micropara-sites and macroparasites, increasing the likelihood of parasite co-occurrence within the same host. This intra-host parasite diversity creates opportunities for one para-site to be a vector for another. Indeed, several ectoparasites transmit pathogens between bats, and bat species that host more ectoparasite species also host greater viral richness [25]. Consequently, if urbanisation alters bat-ectoparasite associa-tions, it could also alter dynamics of pathogen transmission.

One such pathogen of concern is the fungus *Pseudogymnoascus destructans*. This fungus causes white nose syndrome, a disease that mainly kills bats during hibernation and has pushed some North American species to the brink of extinction [26]. Recently, it was detected on spinturnicid mites collected from bats in Kentucky, USA, raising the possibility that ectoparasites are involved in spreading the disease [27]. The fact that these bats were sampled in late summer further suggests that bats might transport the fungus from summer habitats, which may be urban, to their hibernacula, where mating occurs, during which time ectoparasites may move between hosts. Thus, the urban ecology of bat-ectoparasite associations may have conservation implications.

Box 4.1 Ectoparasites of Bats in Relation to Urbanisation
The following case study is extracted from unpublished data from Coleman JL, Swerdfeger E and RMR Barclay.

Problem
Only 29 of the 570 relevant studies identified (Fig. 4.2) assessed bat-parasite relationships along urbanisation gradients, and none did so for colonial insec-tivorous bats in temperate zone cities. As outlined above, examination of bat ectoparasites in urban environments could be key to understanding host roost-ing behaviour and fitness.

Methodology
Ectoparasites on *M. lucifugus* were documented in relation to urbanisation in Calgary, Alberta, Canada. The urban gradient consisted of three zones: urban (within city limits and surrounded by development), rural ($\geq$ 40 km from city limits), and transition (from city limits out to 40 km, 11 sites). Each zone had at least nine replicate sites (11 urban, 11 transition, 9 rural), all located in treed, riparian areas to minimise confounding effects of habitat. From May to mid-September in 2007 and 2008, the authors captured 884 bats by mist-netting and recorded their body condition, demographics, and ectoparasites. The authors considered three cohorts (adult females, adult males, and juveniles) and calcu-lated total ectoparasite prevalence (percentage of bats infested), total

(continued)

Box 4.1 (continued)

ectoparasite intensity (ectoparasites per infested bat), and intensity and prevalence per parasite taxon. Associations between zone and infestation (total ectoparasite prevalence and per parasite taxon) were assessed using two-way contingency tables. The influence of urbanisation on intensity (total ectoparasite intensity and per taxon) was compared for each bat cohort using negative binomial generalised linear mixed models. Finally, non-parametric measures of associations between individual parasite load and body condition were determined.

Findings

The relationship between urbanisation and ectoparasites of *M. lucifugus* reveals a complex response that varies among parasite taxa, over time, and with demographics and body condition. Most bats (60%) harboured at least one ectoparasite (Fig. 4.3), including various mites (Acarina: Macronyssidae, Spinturnicidae), bat fleas (*Myodopsilla* spp., G. Chilton, pers. comm.), bed

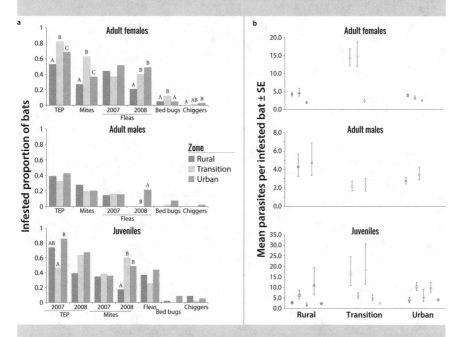

Fig. 4.3 Variation in ectoparasitism on three cohorts of little brown bats, *Myotis lucifugus*, with urbanisation in Calgary, Alberta, Canada. Upper charts for adult females, middle charts for adult males, lower charts for juveniles, different colours for different zones. (**a**) Differences in ectoparasite prevalence (total ectoparasite prevalence = TEP). Different letters above columns indicate significantly different values – columns with no letters are not different. (**b**) Variation in intensity of parasites. Symbols represent total ectoparasites (squares), triangles (mites), and circles (fleas). On the chart for juveniles, closed symbols are 2007 values and open symbols are 2008 values. For adult females and juveniles, values are means over both years. All values are least-squared means with back-transformed standard errors

(continued)

Box 4.1 (continued)

bugs (*Cimex* spp.), chiggers (Trombiculidae, H. Proctor, pers. comm.), and soft ticks (Argasidae). Intensities and yearly variation were generally low, especially in the urban population (Fig. 4.3b). Links between urbanisation and ectoparasitism were obvious in adult female bats. Maximal prevalence (except for fleas in 2007; Fig. 4.3a) and total ectoparasite intensity (Fig. 4.3b) occurred in the transition zone. Total ectoparasite intensity also peaked on lactating females and in 2008. For fleas, intensity differed between years (but not during pregnancy) and with reproductive status in 2008, but not with urbanisation (Fig. 4.3b). Body condition was positively correlated with ecto-parasitism (total ectoparasite intensity and intensities of mites and bed bugs) but only on adult females. For adult male bats, the only link between urban-isation and ectoparasitism was that urban males were the only ones with fleas in 2008 (Fig. 4.3a).

Juvenile bats' ectoparasite associations mirrored those of adult females in some respects. For example, they had higher total ectoparasite and mite inten-sities in 2007, and rural juveniles were less likely to harbour mites in 2008 (Fig. 4.3). Prevalence and intensity measures for other juvenile bat-ectopara-site associations either did not vary with urbanisation or varied inconsistently between years (i.e. significant year-zone interactions).

Synthesis

By some measures, *M. lucifugus* were more parasitised in the city. This is likely because bats were most abundant there [10] and ectoparasitism often rises along with host population density [28]. However, for adult females, parasitism increased in the transition zone. This could reflect divergent roost-ing ecology along the urban-rural gradient. Urban and rural bats mainly roosted in large, enclosed, built structures, while those in the transition zone roosted in tree cavities or under shingles. Though switching among tree roosts can reduce infestations [29], it could also facilitate some dispersal of tempo-rary parasites through passive transport between roosts [30]. Additionally, urbanisation could affect body condition, which was best in the transition zone, and increased parasitism with better body condition is predicted by the hypothesis that parasites prefer healthier hosts [31]. The near lack of variation in ectoparasite associations of adult males along the gradient may simply reflect the fact that they are widely dispersed in summer and harbour few parasites.

For juveniles, low variation in intensities may reflect age-biased parasit-ism. On one hand, the transition zone, where reproductive output peaks [10], presumably offers ectoparasites the greatest availability of young, vulnerable hosts [28]. On the other, because juveniles experience the highest overwinter mortality, permanent parasites should avoid independent young prior to win-ter [11] regardless of urbanisation.

Box 4.1 (continued)

Overall, this case demonstrates that ectoparasitic associations and their links to urbanisation can vary widely among conspecific cohorts. This highlights how the ecology of colonial bats can differ within as well as among species. It also illustrates that short-term studies might not reveal the full picture of urbanisation-mediated ectoparasitism. In some ways, parasitism did differ between years but it was most consistent in the city. This could reflect the potential for reduced urban seasonality (i.e. UHI) affecting parasites directly, by influencing their survival or, indirectly, by influencing host population dynamics and movements [28]. Ultimately, this case underscores the importance of multi-year investigations of multiple infestation metrics and parasitic taxa along urbanisation gradients to elucidate the role of urbanisation in mediating bat-ectoparasite relationships.

3 Endoparasites

As human encroachment on bat habitat grows, so do worries about bats acting as reservoirs for parasites, including some of public health concern. However, bat parasites that do not infect humans or domestic animals can provide key clues to ecological differences between urban and non-urban bats, such as feeding and roosting preferences. Additionally, comparing parasite diversity between urban and non-urban bats can help elucidate whether key phenomena, such as biological homogenisation, occur at multiple scales within anthropogenically disturbed habitat. Eukaryotic bat endoparasites, typically single-celled protozoans and worm-like helminths, represent both tropically transmitted and vector-borne groups. Thus, these parasites can reveal the influence of land use and land cover change on host susceptibility, parasite contact rates, and transmission pathways and the potential consequences of land use and land cover change on biodiversity at the scales of the host and parasite.

3.1 Protozoan Parasites and Host-Vector Contact Rates

Blood-borne parasites in the genera *Trypanosoma* and *Leishmania* (phylum Euglenozoa) are transmitted by hematophagous insects. Although both New and Old World *Leishmania* spp. have been documented infecting bats [e.g. 32 and others therein], only New World *Trypanosoma* spp. have been found in bats [33, 34]. One such species, *T. cruzi*, which causes Chagas disease in humans, infects various mammals, and there are concerns that bats could act as reservoirs of this parasite. Indeed, the prevalence of *Trypanosoma* spp. infecting Jamaican fruit-eating bats (*Artibeus jamaicensis*) is higher in forest fragments in a residential and agricultural matrix than in continuous tropical forest in Panama [34]. Likewise, trypanosomes were isolated from five bat species within rainforest fragments and surrounding farms in Espirito Santo, Brazil, but not from 20 other species of wild mammals [33]. The prevalence of *Leishmania* spp. within urban and peri-urban bats may be relatively high. For example, over 59% of common pipistrelle (*Pipistrellus pipistrellus*) sampled in and around Madrid harboured *L. infantum* [32], whereas *Miniopterus schreibersii* in Spanish wildlands demonstrated no evidence of *Leishmania* infection [35]. Although the dichotomous findings of these studies could reflect differences in tissues examined (i.e. the spleen [32] versus peripheral blood [35]), human-modified landscapes provide phlebotomine sandflies, the vectors of *Leishmania* spp., with hospitable habitat [36], and these flies can feed successfully on multiple species of bats [37]. Thus, an increase in phlebotomine sandflies in urban areas could be responsible for higher urban infection rates. Given that both parasite genera (*Leishmania* and *Trypanosoma*) have generalist species and generalist arthropod vectors, urbanisation could increase parasite contact rates for urban bats.

Members of the phylum Apicomplexa parasitise a wide variety of birds and mammals, including bats, and some are of zoonotic concern [38]. Apicomplexans can enter hosts via a hematophagous arthropod vector (e.g. *Plasmodium*) or through faecal-oral transmission (e.g. *Eimeria*). Therefore, effects of urbanisation in this phylum could vary with the life cycle and vector. For instance, Indian flying foxes (*Pteropus medius*) were slightly more likely to host *Hepatocystis* sp., vectored by mosquitos, and *Babesia* sp., vectored by ticks, in peri-urban than in rural areas of Bangladesh [39]. Meanwhile, the prevalence of *Polychromophilus* sp., vectored by bat flies, in Australian bent-wing bats (*Miniopterus orianae*) was up to 1.9 times higher at sites that retained ≤18% of their original habitat than at sites with ≥45% [40]. Thus, land use changes may promote apicomplexan infections, perhaps by increasing vector-host contact rates and/or susceptibility of hosts.

3.2   Helminths Provide Insights into Host Foraging

Although roundworms (Nematoda) and spiny-headed worms (Acanthocephala) infect bats, flukes (Trematoda) and tapeworms (Cestoda) often dominate bat helminth communities [41]. Many of these parasites have complex life cycles involving one or more invertebrate intermediate hosts. Whereas bat trematodes require two aquatic intermediate hosts (freshwater snails and larval insects; Fig. 4.4), cestodes have fully terrestrial life cycles, with arthropods, e.g. beetles, acting as single intermediate hosts [42]. Thus, habitat diversity of bat helminth life cycles varies, and anthropogenic disruption of any of these habitats may shift helminth communities. Urbanisation can affect both parasite community diversity and host traits, e.g. body condition and immune function. It could also cause ecologically sensitive intermediate hosts to decline, while more resilient taxa could become dominant [43]. Similarly, certain urban stressors could increase host susceptibility via

Fig. 4.4 Representative life cycle diagram of a trematode belonging to Lecithodendriidae, a family that almost exclusively parasitises bats. Trematode eggs are passed with faeces as a bat flies over a body of freshwater, such as when drinking (**a**). They then infect a snail where larvae metamorphose, grow, and exit their host as another free-swimming larval stage (**b**). These larvae swim until they contact a larval insect (e.g. a dragonfly nymph) and encyst within it (**c**). When the dragonfly metamorphoses into an adult, it carries the encysted trematode larvae (**d**). When a bat ingests the adult dragonfly, the encysted larvae break free and grow into adult worms in the bat's intestine, where they begin shedding eggs with the bat's faeces (**e**)

physiological processes [44]. Ultimately, understanding urban-related shifts in helminth communities may reveal the responses of bat hosts to extreme habitat disturbance.

Box 4.2 The Link Between Bat Helminth Communities and Anthropogenic Land Use
This information is extracted from Warburton et al. (2016) [41].

Problem
The ecology of endoparasite communities that inhabit bats is understudied not only in relatively undisturbed settings but also in relation to urbanisation. These helminth communities exist across a variety of environmental conditions, including not only 'natural' but also highly altered land covers, such as urbanised ones. Biological communities are typically thought to exhibit a distance-decay relationship where their species compositions become increasingly dissimilar with increasing physical distance. However, environments

(continued)

Box 4.2 (continued)

themselves, especially anthropogenically altered ones, can also shift the species composition of biological communities. Further, understanding how parasite communities change with urbanisation may elucidate how bat hosts function in cities by revealing key aspects of bat foraging ecology in urban environments.

Methodology

To understand how anthropogenic habitat disturbance contributes to shifting parasite communities, *Eptesicus fuscus* from a three-state region (Michigan, Indiana, and Kentucky) in the Midwestern USA were captured, and their helminth communities were assessed. Two hundred sixty bats consisting of adult and juvenile members of both sexes were captured from 13 maternity colonies with a mean inter-roost distance of 315.7 km (range = 6.9–660.7 km). The authors used GIS layers from the US National Land Cover Database and National Wetlands Inventory to quantify the area covered by 16 land cover categories, including designations such as barren land, croplands, forests, wetlands, and city centres, within 12-km radii of each colony (i.e. the maximum recorded foraging distance for *E. fuscus*). Using redundancy analysis, an extension of multiple linear regression that accounts for multiple response and explanatory variables, the effects of physical distances between roosts and land cover on helminth communities were assessed.

Findings

Helminth community composition was largely predicted by land cover around roosts. Indeed, land cover categories with more intense human activities had similar helminth communities. The effect was most significant ($p < 0.004$) in developed open spaces (e.g. parks, golf courses) and high-impervious cover sites (e.g. central business districts) and approached significance ($p = 0.0504$) in cultivated land covers (e.g. croplands, orchards). However, more urbanised sites did not have less species rich or less diverse helminth communities; instead, their species composition changed. Certain helminths, e.g. the cestode *Hymenolepis roudabushi* and the trematode *Paralecithodendrium swansoni*, were more closely associated with more developed land cover, while others, e.g. the nematodes *Rictularia lucifugus* and *Litomosoides guitaresi*, were more closely associated with cropland. Still other species, e.g. the trematode *Acanthatrium eptesici*, were associated with relatively undisturbed habitats such as woody wetlands.

Synthesis

These shifts in the helminth communities of bats in different land covers likely reflect shifts in intermediate host community composition and structure. Parasites with ecologically sensitive intermediate hosts, such as mayflies, might be more prone to extirpation with increasing anthropogenic

(continued)

Box 4.2 (continued)

disturbance. However, instead of producing a net loss in parasite species richness, the ecological niches left vacant by such extirpations could be filled by other helminths whose intermediate hosts are more resilient, such as chrysomelid beetles. Thus, understanding how helminth communities change in urban areas can reveal bats use resources in anthropogenically altered landscapes.

4 Concluding Perspectives

After surveying the literature, there is clearly still much to learn about bat-parasite relationships in the context of urbanisation. This knowledge gap is unfortunate, although perhaps not surprising given that bats and parasites are high-diversity groups, and they occur in many cities around the world. Further, the tightly linked nature of the host-parasite relationship provides excellent opportunities to address key urban ecology questions.

Questions about host foraging and roosting habits in urban areas, host-vector contact rates in disturbed habitat, and host susceptibility in response to anthropogenic stressors can be readily addressed within urban bat-parasite systems. For example, many ectoparasites contact bat hosts in roosts, whereas many helminths of bats are tropically transmitted. As such, comparing the diversity of ectoparasite and endoparasite communities between urban and non-urban bats can provide insight into how urban bats use resources in response to anthropogenic disturbance.

Additionally, certain human activities could increase transmission pathways, but this phenomenon is largely unexamined for most parasitic taxa. In one well-known example [5], urban planting of ornamental trees increased aggregations of flying foxes and consequently led to increased HeV transmission. Given that HeV relies on faecal-oral transmission, parasites with faecal-oral transmission (e.g. coccidia) could increase in these cities as well. Other human activities, such as draining wetlands for residential or agricultural use, should eliminate transmission pathways for trematodes that use aquatic intermediate hosts. However, anthropogenic effects on transmission pathways are poorly studied for most parasitic taxa, including those parasitising bats, and represent key knowledge gaps that require further investigation.

Urban bat-parasite systems could also be useful for examining broader ecological hypotheses. One is the diversity dilution hypothesis, which predicts increasing parasitism with declining diversity of hosts. Although evidence is equivocal [45], some findings in anthropogenically disturbed habitats [46–48] support key aspects of the hypothesis, namely, that preserving biodiversity in urban areas can reduce disease incidence.

Because parasitic associations are strong selective forces on both partners [14], urbanisation could have evolutionary implications for hosts and parasites. For example, pigeons (*Columbia livia*) exhibit hereditary variation in colouration along the urban gradient in Paris, France – variation that apparently reflects divergent strategies to cope with urban-related changes in blood-parasite pressure [49]. Although urban evolutionary ecology studies have not yet focused on bats and their parasites, doing so could elucidate the role of cities as drivers of evolution.

Research on urban bats and their parasites could also have important ecotoxicological applications as diverse parasites are increasingly perceived as useful bioindicators of habitat quality [50]. Finally, the potential effects of anthropogenic stressors, such as light pollution and roost disturbance, on the immune system of urban bats are not well known. These stressors could have a negative impact on disease susceptibility in urban bats, thereby increasing parasite prevalence or abundance. Thus, future work linking environmental health, anthropogenic activities, and host susceptibility could shed more light on our understanding of urban bat-parasite systems.

Literature Cited

1. Zelmer DA (1998) An evolutionary definition of parasitism. Int J Parasitol 28(3):531–533
2. Frick WF, Kingston T, Flanders J (2020) A review of the major threats and challenges to global bat conservation. Ann N Y Acad Sci 1469(1):5–25
3. Murray MH et al (2019) City sicker? A meta-analysis of wildlife health and urbanization. Front Ecol Environ 17(10):575–583
4. Williams NSG et al (2006) Range expansion due to urbanization: increased food resources attract grey-headed flying-foxes (Pteropus poliocephalus) to Melbourne. Austral Ecol 31(2):190–198
5. Paez DJ et al (2018) Optimal foraging in seasonal environments: implications for residency of Australian flying foxes in food-subsidized urban landscapes. Philos Trans R Soc B, Biol Sci 373(1745)
6. Plowright RK et al (2011) Urban habituation, ecological connectivity and epidemic dampening: the emergence of Hendra virus from flying foxes (Pteropus spp.). Proc R Soc B Biol Sci 278(1725):3703–3712
7. Webber QMR, Willis CKR (2016) Sociality, parasites, and pathogens in bats. In: Ortega J (ed) Sociality in bats. Springer International Publishing, Cham, pp 105–139
8. Russo D, Ancillotto L (2015) Sensitivity of bats to urbanization: a review. Mamm Biol 80(3):205–212
9. Ancillotto L, Tomassini A, Russo D (2016) The fancy city life: Kuhl's pipistrelle, Pipistrellus kuhlii, benefits from urbanisation. Wildl Res 42(7):598–606
10. Coleman JL, Barclay RMR (2011) Influence of urbanization on demography of little brown bats (Myotis lucifugus) in the prairies of North America. PLoS One 6(5):e20483
11. Zahn A, Rupp D (2004) Ectoparasite load in European vespertilionid bats. J Zool 262(4):383–391
12. Korine C et al (2017) The effect of water contamination and host-related factors on ectoparasite load in an insectivorous bat. Parasitol Res 116(9):2517–2526
13. Serieys LEK et al (1871) Urbanization and anticoagulant poisons promote immune dysfunction in bobcats. Proc R Soc B Biol Sci 2018(285):20172533
14. Marshall AG (1982) Ecology of insects Ectoparasitic on bats. In: Kunz TH (ed) Ecology of bats. Plenum Publishing Corporation, New York, pp 369–401

15. Nunes H, Rocha FL, Cordeiro-Estrela P (2017) Bats in urban areas of Brazil: roosts, food resources and parasites in disturbed environments. Urban Ecosyst 20(4):953–969
16. Bergeson SM, Holmes JB, O'Keefe JM (2020) Indiana bat roosting behavior differs between urban and rural landscapes. Urban Ecosyst 23(1):79–91
17. Lim ZX et al (2020) Ecology of bat flies in Singapore: a study on the diversity, infestation bias and host specificity (Diptera: Nycteribiidae). Int J Parasitol Parasites Wildl 12:29–33
18. Webber QMR, Czenze ZJ, Willis CKR (2015) Host demographic predicts ectoparasite dynamics for a colonial host during pre-hibernation mating. Parasitology 142(10):1260–1269
19. Lourenço S, Palmeirim JM (2008) Which factors regulate the reproduction of ectoparasites of temperate-zone cave-dwelling bats? Parasitol Res 104(1):127
20. Pilosof S et al (2012) Effects of anthropogenic disturbance and climate on patterns of bat fly parasitism. PLoS One 7(7):e41487
21. Buczek A et al (2014) Threat of attacks of Ixodes ricinus ticks (Ixodida: Ixodidae) and Lyme borreliosis within urban heat islands in South-Western Poland. Parasit Vectors 7(1):562
22. Ruiz SR et al (2019) Metal and metalloid exposure and oxidative status in free-living individuals of Myotis daubentonii. Ecotoxicol Environ Saf 169:93–102
23. McMahon TA, Rohr JR, Bernal XE (2017) Light and noise pollution interact to disrupt interspecific interactions. Ecology 98(5):1290–1299
24. Mollentze N, Streicker DG (2020) Viral zoonotic risk is homogenous among taxonomic orders of mammalian and avian reservoir hosts. Proc Natl Acad Sci 117(17):9423
25. Gay N et al (2014) Parasite and viral species richness of southeast Asian bats: fragmentation of area distribution matters. Int J Parasitol Parasites Wildl 3(2):161–170
26. Cheng TL et al (2021) The scope and severity of white-nose syndrome on hibernating bats in North America. Conserv Biol
27. Mustachio A, Bodri MS (2019) Can ectoparasites be implicated in the spread of Pseudogymnoascus destructans? J Wildl Dis 55(3):704–706
28. Bradley CA, Altizer S (2007) Urbanization and the ecology of wildlife diseases. Trends Ecol Evol 22(2):95–102
29. Lewis SE (1995) Roost fidelity of bats: a review. J Mammal 76(2):481–496
30. Reinhardt K, Siva-Jothy MT (2007) Biology of the bed bugs (Cimicidae). Annu Rev Entomol 52(1):351–374
31. Christe P et al (2003) Differential species-specific ectoparasitic mite intensities in two intimately coexisting sibling bat species: resource-mediated host attractiveness or parasite specialization? J Anim Ecol 72(5):866–872
32. Azami-Conesa I et al (2020) First detection of Leishmania infantum in common urban bats Pipistrellus pipistrellus in Europe. Res Vet Sci 132:172–176
33. Acosta IDCL et al (2014) Survey of Trypanosoma and Leishmania in wild and domestic animals in an Atlantic rainforest fragment and surroundings in the state of Espírito Santo. Braz J Med Entomol 51(3):686–693
34. Cottontail VM, Wellinghausen N, Kalko EKV (2009) Habitat fragmentation and haemoparasites in the common fruit bat, Artibeus jamaicensis (Phyllostomidae) in a tropical lowland forest in Panamá. Parasitology 136(10):1133–1145
35. Millán J et al (2014) Absence of Leishmania infantum in cave bats in an endemic area in Spain. Parasitol Res 113(5):1993–1995
36. Rosário ING et al (2016) Evaluating the adaptation process of sandfly fauna to anthropized environments in a leishmaniasis transmission area in the Brazilian Amazon. J Med Entomol 54(2):450–459
37. Lampo M et al (2000) A possible role of bats as a blood source for the Leishmania vector Lutzomyia longipalpis (Diptera: Psychodidae). AJTHAB 62(6):718–719
38. Schiller SE, Webster KN, Power M (2016) Detection of Cryptosporidium hominis and novel cryptosporidium bat genotypes in wild and captive Pteropus hosts in Australia. Infect Genet Evol 44:254–260

39. Islam S et al (2020) Detection of hemoparasites in bats, Bangladesh. J Threat Taxa 12(10):16245–16250
40. Holz PH et al (2019) Polychromophilus melanipherus and haemoplasma infections not associated with clinical signs in southern bent-winged bats (Miniopterus orianae bassanii) and eastern bent-winged bats (Miniopterus orianae oceanensis). Int J Parasitol Parasites Wildl 8:10–18
41. Warburton EM, Kohler SL, Vonhof MJ (2016) Patterns of parasite community dissimilarity: the significant role of land use and lack of distance-decay in a bat–helminth system. Oikos 125(3):374–385
42. Etges FJ (1960) On the life history of Prosthodendrium (Acanthatrium) anaplocami n. sp. (Trematoda: Lecithodendriidae). J Parasitol 46(2):235–240
43. Mykrä H, Heino J, Muotka T (2007) Scale-related patterns in the spatial and environmental components of stream macroinvertebrate assemblage variation. Glob Ecol Biogeogr 16(2):149–159
44. Allen LC et al (2009) Roosting ecology and variation in adaptive and innate immune system function in the Brazilian free-tailed bat (Tadarida brasiliensis). J Comp Physiol B: Biochem Syst Environ Physiol 179(3):315–323
45. Rohr JR et al (2020) Towards common ground in the biodiversity–disease debate. Nat Ecol Evol 4(1):24–33
46. LoGiudice K et al (2003) The ecology of infectious disease: effects of host diversity and community composition on Lyme disease risk. Proc Natl Acad Sci 100(2):567–571
47. Rubio AV, Ávila-Flores R, Suzán G (2014) Responses of small mammals to habitat fragmentation: epidemiological considerations for rodent-borne hantaviruses in the Americas. EcoHealth 11(4):526–533
48. Ezenwa VO et al (2007) Land cover variation and West Nile virus prevalence: patterns, processes, and implications for disease control. Vector-Borne Zoonotic Dis 7(2):173–180
49. Jacquin L et al (2013) Melanin-based coloration is related to parasite intensity and cellular immune response in an urban free living bird: the feral pigeon Columba livia. J Avian Biol 42(1):11–15
50. Sures B et al (2017) Parasite responses to pollution: what we know and where we go in 'environmental parasitology'. Parasit Vectors 10(1):65

Chapter 5
Bat Societies across Habitat Types: Insights from a Commonly Occurring Fruit Bat *Cynopterus sphinx*

Kritika M. Garg, Balaji Chattopadhyay, D. Paramanatha Swami Doss, A. K. Vinoth Kumar, and Sripathi Kandula

Abstract Bats constitute the second most speciose order of mammals and are known for their gregarious and flexible social structure. Urbanisation can lead to changes in the availability of, or access to, resources, such as roosting sites, food, and mates, and thus can potentially affect the social and mating systems. In this chapter, we summarise knowledge of the effects of urbanisation on the social structure of bats and highlight gaps in the literature. Further, we discuss the social structure and reproductive output of *Cynopterus sphinx*, a fruit bat ubiquitous across human-dominated habitats in South and Southeast Asia. We followed two *C. sphinx* colonies over multiple seasons to understand how urbanisation impacts the social systems of these colonies. We used direct observational and genetic data to compare the colony size, social subunit size, relatedness, and reproductive output of both colonies. On average, the rural colony was larger than the urban colony. The two

K. M. Garg (✉)
Department of Biology, Ashoka University, Sonipat, India

Centre for Interdisciplinary Archaeological Research, Ashoka University, Sonipat, India
e-mail: kritika.garg@ashoka.edu.in

B. Chattopadhyay
Trivedi School of Biosciences, Ashoka University, Sonipat, India

School of Biological Sciences, Madurai Kamaraj University, Madurai, India

D. Paramanatha Swami Doss
School of Biological Sciences, Madurai Kamaraj University, Madurai, India

St. John's College, Palayamkottai, India

A. K. Vinoth Kumar
School of Biological Sciences, Madurai Kamaraj University, Madurai, India

S. Kandula
School of Biological Sciences, Madurai Kamaraj University, Madurai, India

Faculty of Allied Health Sciences, Chettinad Academy of Research and Education, Chettinad Health City, India

L. Moretto et al. (eds.), *Urban Bats*, Fascinating Life Sciences,
https://doi.org/10.1007/978-3-031-13173-8_5

colonies did not differ significantly with respect to social subunit size and related-ness, suggesting minimal impact of urbanisation on social structure. However, there was a significant difference in reproductive output, with the reproductive success of females from the rural colony being 1.7 times greater than that of the urban colony. Our results from a single rural and urban colony located nearly 500 km apart suggest that urbanisation may reduce fecundity and urban areas may act as ecological traps. Future studies with more extensive sampling are needed to identify the main drivers of female reproductive success and the cause of reduced female fecundity in urbanised habitats.

Keywords Urbanisation · Reproductive success · Social system · India

1 Introduction

Humans have had a drastic impact on the natural environment, altering most of the terrestrial landscape and causing major loss of suitable habitat and biodiversity [1–4]. While human activities have profoundly affected the planet since the late Pleistocene [1], conversion of large tracts of forests to agricultural lands in the past century has caused an unprecedented level of environmental degradation and the sixth mass extinction [1, 2]. However, it is only now that we are beginning to under-stand and appreciate the impact of urbanisation on animal societies [3–5].

A social system is the relationship between conspecifics of a group, where indi-viduals within a group interact more with each other than with conspecifics from another group [6]. Such interactions are in turn influenced by the individuals' envi-ronment [6], including distribution of resources, predation risk, and access to mates [6–8]. The distribution and abundance of resources in a given area directly affect how many individuals are supported, group size, dispersal rates, and mating oppor-tunities, and thus ultimately affect the social system, social structure, and mating system [5, 7–9]. Urbanisation leads to habitat fragmentation, altered predation pres-sure, and an imbalance in resource availability, thereby potentially affecting social and mating systems [1–4, 9]. Throughout this chapter, we follow Kappeler and van Schaik's [6] definition of social system and social structure, wherein social system consists of the social organisation, mating system, and social structure. Social organisation refers to the group composition (size and sex ratio) of the social group. Social structure refers to the social interactions and relationships among individu-als. Mating system encompasses behavioural and genetic components of mat-ing [6–8].

Interestingly, bats are among the most common mammals inhabiting urbanised landscapes [4, 10]. They are keystone species in many ecosystems and particularly sensitive to environmental disturbance, making them good indicators of ecosystem quality [11]. Most bats are gregarious and exhibit flexible mating and social systems [10, 12, 13]. However, knowledge of the impact of urbanisation on social systems of the more than 1400 bat species known to date is limited.

Roosting sites are key resources for bats and [4, 10, 14] play an important role in shelter, protection from predators, rearing of offspring, and social interactions [4, 10, 14]. In many species, males defend these roost sites, resulting in uneven access to females and a polygynous mating system [12, 14]. Urban landscapes may provide ample roosting sites for bats that are specialised for roosting in cliff crevices [4]. Bats can occupy tunnels, bridges, and crevices in buildings. But not all species can use these structures, especially cave- and foliage-roosting species, and these species might be negatively affected by urbanisation if required resources for these species are not available in urban areas [4, 10, 14].

Urbanisation can also bring about a change in social organisation [5, 9]. Due to limited resources, sex-biased difference in exploitation of urban habitats is observed in many species [4, 5, 9, 10, 15–27]. For example, Linott et al. found that, in soprano pipistrelles (*Pipistrellus pygmaeus*), differential habitat use between males and females has been observed [15]. Females were less likely than males to use poorly connected woodlands with few mature trees. Further, females were most likely to be captured in woodlands surrounded by waterways. In contrast, no preferential habitat use by males was observed [15], and male capture rates did not vary with habitat quality [15]. However, the impact of poor habitat quality on male reproductive success is yet to be ascertained [15]. Similarly, in many insectivorous bats, due to high-energy demands, lactating females prefer better quality habitats compared to males, especially habitats associated with water bodies [15, 17]. Thus, urbanisation can affect a population's sex ratio, social organisation, and, ultimately, its social system.

Some bats can effectively occupy urban habitats and are suspected to thrive in urban environments. However, detailed long-term studies are required to assess if urban areas provide suitable habitat for multiple bat species. One of the pioneering studies quantifying the effect of urbanisation on the synanthropic (species which are common and may benefit from living in urban areas) little brown bat (*Myotis lucifugus*) observed that bats in the transition habitat between rural and urban areas had the highest reproductive success and their body condition was better than that of urban and rural bats [18]. Conversely, higher reproductive output and juvenile fledging rate were observed in urban than in non-urban populations of Kuhl's pipistrelle (*P. kuhlii*), suggesting the need for detailed studies on the impact of urbanisation in multiple species [19]. In this chapter, we summarise current knowledge of the impact of urbanisation on social and mating systems of a commonly occurring fruit bat and highlight areas of future research.

2   Societies of the Short-Nosed Fruit Bat *Cynopterus sphinx* Across Urban and Rural Landscapes

We investigated the impact of urbanisation on the synanthropic fruit bat *C. sphinx*, a generalist species found across South and Southeast Asia [20]. Males of this species modify foliage to construct their own tents, which they use as roosts [20–22].

Individuals also use human-made structures as roosts, allowing them to exploit a variety of habitat types [20, 23]. Given that the species is omnipresent in human-occupied landscapes and can use artificial structures for its benefit, this is an ideal system to understand the impacts of urbanisation.

Typically, once the male constructs a tent, females join the tent and reside with the male, suggestive of a harem-like social and mating unit [20, 22, 23]. Although the species has two reproductive periods (February to March and October to November), harems are observed throughout the year [20]. Long-term behavioural and genetic studies on this species have demonstrated that despite this species' harem-like social organisation, its social structure and mating system do not truly conform to a harem system [22–24]. Rather, group composition changes on a regular basis; with females moving freely between groups [22–24], the mating system is promiscuous, and there is no correlation between harem size and male reproductive success [22]. Males in a colony enjoy greater reproductive success compared to solitary males [22]. Thus, the term harem is a misnomer in this species, and for the rest of the chapter, we use the term "social subunit" instead [23, 24]. In this species, multiple subunits along with solitary males that are clustered in an area are collectively termed as a "colony", which constitutes the main social and reproductive unit [20, 22, 23].

In this study, we compared the social organisation of two colonies of *C. sphinx*, one each from an urban and rural habitat in India. The urban population was located in the metropolitan city of Bangalore, at the Indian Institute of Science campus (12.99° N, 77.59° E). *C. sphinx* occupied both human-made (window eaves) and natural roosts (kitul palm, *Caryota urens* fruiting body, and ornamental palm leaves) at this site [23]. The rural population was located approximately 500 km away at Samyathu village (8.638° N, 77.958° E), in the state of Tamil Nadu, and the population only roosted in leaves of palmyra palm, *Borassus flabellifer* [22, 24]. The rural colony was surrounded by farmland, and banana is regularly cultivated in the region [22, 24]. We used long-term behavioural observations and genetic data to compare the social organisation and reproductive output between both colonies. We specifically asked if there were any differences in colony size, relatedness levels, and reproductive output. We predicted that the rural colony would be larger than the urban colony because the rural habitat can support more individuals due to the presence of higher resources. Further, behavioural observations suggest similar colony composition of both rural and urban colonies. Therefore, we expected no differences in social organisation between rural and urban colonies.

The rural population was part of a long-term study that we sampled twice (to capture both reproductive periods) every year from 2008 to 2012 ($N = 10$ observations) [22, 24]. We sampled the urban population for three consecutive reproductive seasons, from August 2011 to 2013 ($N = 3$ observations) [23]. Our sampling timing was such that most of the females had already given birth, but the pups were not yet weaned [22–24]. This common protocol ensured that we could assign most pups to their mothers based on their associations and test these associations using genetic markers. Generally, we sampled both colonies approximately 4 weeks after observing parturition (through the visual inspection of newborns attached to adult females).

We carried out a visual census of the entire colony prior to sampling. We used hoop nets to capture the social subunit and sampled entire colonies, noting any escapees. We kept bats in cloth bags prior to sampling. For each individual, we measured its forearm length and body mass and noted its sex, age, and reproductive status [22–24]. We collected a 6-mm/4-mm sterile biopsy punch from both wings for genetic analysis. We also tagged all captured individuals using colour-coded bead necklaces.

We captured a total of 396 adults across 10 sampling periods at the rural colony and 81 adults across 3 sampling periods at the urban colony [22–24]. We extracted DNA using either modified salt chloroform extraction protocol or Qiagen DNeasy Blood and Tissue Kit. We genotyped all samples for nine microsatellite loci and checked all genotypes manually twice to test for consistency. We removed one locus from all analyses due to the presence of null alleles [22–24]. For further details on microsatellite typing and subsequent processing of genetic data, see Garg et al. [22].

Our analyses revealed that the rural colony (only considering adults) was slightly larger (mean = 40 adults ± 14 SD; N = 10 observations) than the urban colony (mean = 32 adults ± 7 SD; N = 3 observations). The rural colony had nearly twice as many social subunits (11.1 ± 2.6 SD) as the urban colony did (6.3 ± 2.1 SD). Although the average social subunit size was slightly smaller in the rural colony (mean = 2.48) than in the urban colony (mean = 3.22), social subunit size did not differ significantly between the two (117 rural subunits, 18 urban subunits, Wilcoxon rank sum test $W = 925.5$, $P = 0.39$; Fig. 5.1).

We further computed each colony's reproductive output, which we defined as the number of offspring captured from a social subunit divided by the number of adult females captured from that subunit. Our observations revealed a significant difference in the reproductive output of females, where 78% of adult females from the rural colony gave birth to at least one offspring compared to 45% of adult females from the urban colony (number of observations for the rural colony, 117; number of observations for the urban colony, 18; Wilcoxon rank sum test $W = 1491$, $P = 0.002$; Fig. 5.2). Overall, to test if differential body condition explained the difference in

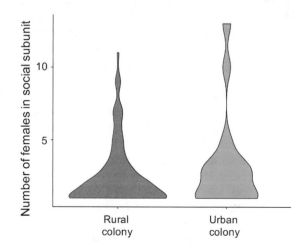

Fig. 5.1 Violin plot depicting the variation in the number of females associated with a male in a social subunit across rural and urban colonies of *Cynopterus sphinx*

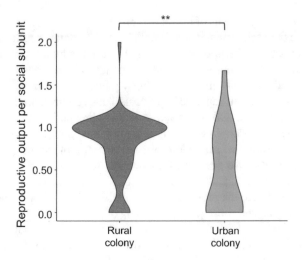

Fig. 5.2 Violin plot depicting the significant difference in reproductive output of rural and urban colonies of *Cynopterus sphinx* (number of subunits from the rural colony, 114; number of subunits from the urban colony, 20; Wilcoxon rank sum test *p* value = 5.503E−05). Reproductive output is quantified as the ratio of the number of offspring captured from a social subunit to the number of adult females captured from that social subunit. ** denotes *p* value <0.01

reproductive output, we further compared forearm length, body mass, and ratio of body mass to forearm length of females from both colonies. Our data revealed that females from the rural colony were significantly smaller than females from the urban colony ($N = 283$ rural females, 50 urban adult females, Welch two sample t-test, $P = 1E-07$; body mass, Wilcoxon rank sum test $P = 4.693E-06$; ratio of body mass to forearm length, Wilcoxon rank sum test $W = 5064.5$, $P = 0.0014$; Fig. 5.3). This was unexpected because given the difference in reproductive output, the body condition of bats should be worse in the urban than in the rural colony. However, given that clinal variation is observed in *Cynopterus sphinx* and bats from the southern latitudes are smaller than bats from the northern latitude [26–28], body size/ condition may not be a good predictor of female reproductive fitness in this fruit bat. Even after controlling for clinal variation by comparing the ratio of body mass to forearm length, we observed that rural females were smaller than urban females (Fig. 5.3c). Future studies must control for latitudinal effects to understand the biological reason for low reproductive output of urban colonies.

Finally, we compared the genetic relatedness between colony members. We estimated pairwise Queller-Goodnight relatedness [29] between adult individuals within a colony in COANCESTRY v 1.0.1.7 software [30]. Although the social and mating implications remain unclear, relatedness plays an important role in the formation of *C. sphinx* social subunits, wherein the male is related to one of the females in the social subunit irrespective of the time of sampling [31]. Therefore, we also investigated average relatedness of the social subunit. For all social subunits where there were no escapees during capture, we calculated pairwise relatedness between all adults within the social subunit. We tested whether a social subunit's average

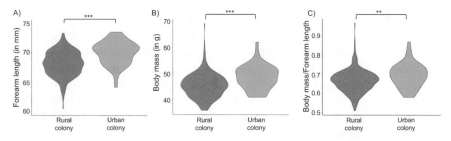

Fig. 5.3 Violin plot depicting the significant difference in (**a**) forearm length, (**b**) body mass, and (**c**) ratio of body mass to forearm length of adult females of *Cynopterus sphinx* across rural and urban colonies (number of females captured from the rural colony, 283; number of females captured from the urban colony, 50; forearm length, Welch two sample *t*-test, *p* value = 1E−07; body mass, Wilcoxon rank sum test *p* value = 4.693E−06). **denotes *p* value <0.01, ***denotes *p* value <0.001

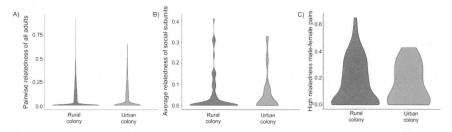

Fig. 5.4 Violin plot depicting the distribution of (**a**) pairwise relatedness between adults from rural and urban colonies, (**b**) average relatedness of social subunits from rural and urban colonies, and (**c**) high relatedness of male-female pairs form each social subunit from urban and rural colonies

relatedness differs significantly between rural and urban colonies. Further, we selected the male-female pair within a social subunit that had the highest relatedness values and tested whether this value differed between both colonies. We discarded the rare observations (three in the rural colony, one in the urban colony) of two adult males in the same social subunit from our analysis of social subunit relatedness. Overall relatedness among the colony members was low, and we did not find any significant correlation between long-term association among colony members and genetic relatedness [23, 24]. We also observed no significant difference in pairwise relatedness between both colonies, suggesting that social organisation of *C. sphinx* does not differ drastically (*N* = 78,934 and 2775 pairwise observations for the rural and urban colonies, respectively; Wilcoxon rank sum test *W* = 110,770,000, *P* = 0.27; Fig. 5.4a). There was also no significant difference in social subunit formation in both colonies. The average relatedness of the social subunit in both colonies was similar (*N* = 78 rural and 15 urban social subunits; Wilcoxon rank sum test *W* = 588.5, *P* = 0.97; Fig. 5.4b). Even the highly related male-female pairs from each social subunit exhibited similar relatedness values across both urban and rural colonies (*N* = 78 rural and 15 urban social subunits; Wilcoxon rank sum test *W* = 585, *P* = 1; Fig. 5.4c).

3   Discussion

Urbanisation is a strong selective force that is currently shaping the biodiversity of our planet [3, 4]. As urban areas continue to expand, many species face rapid decline and even local extinction. We are now in the Anthropocene epoch, characterised in part by the sixth mass extinction, so there is a pressing need to prevent further declines and conserve our current biodiversity [32]. Species biology plays an important role in determining the impacts of urbanisation, with many bat species exhibiting negative responses to urbanisation and only a handful of them being able to occupy the new "urban niche" that is available to them [4, 9, 10]. Research on the impacts of individual interactions, behaviour, and mating is still in its infancy, and these impacts remain more difficult to quantify than species abundance and genetic connectivity. However, multiple research groups are making strides in this area, and we are beginning to quantify the life history differences between urban and non-urban populations, which are of paramount importance for effective conservation planning.

3.1   *Potential Impact of Urbanisation on* Cynopterus sphinx

In the present study, we did not observe any major difference in the social organisation, social structure, social subunit size, and genetic relatedness of *C. sphinx* between the rural and urban colonies (Figs. 5.1 and 5.4). However, the size of social subunits in the urban colony was nearly half that of the rural colony, and there was a dramatic decline in the reproductive output of the females in the urban colony (Fig. 5.2). Body condition did not appear to influence female fecundity within our urban *C. sphinx* colony, and future studies will hopefully shed more light on why reproductive output was lower. It should be noted that the current study is based on data collected from two colonies separated by hundreds of kilometres. The lack of replication and possible latitudinal effects (e.g. temperature, precipitation, prey availability) may explain differences in reproductive success independent of urbanisation. Differences in access to mates may be involved as well. Thus, to clearly determine if in fact urbanisation is influencing populations of *C. sphinx*, future studies must also incorporate data from multiple colonies across a habitat gradient and control for latitudinal variation.

Future studies must also consider species-specific and sex-specific differences in resource requirements while studying the impact of urbanisation. Resource distribution and availability have a strong impact on the distribution of individuals, social organisation, mating system, and ultimately social system [7]. Most bats are gregarious in nature, and flight allows them to traverse greater distances than other terrestrial species of similar body size [12–14]. However, this does not make them immune to the effects of urbanisation. The impact of urbanisation on bats is species-specific because bats have different roosting requirements and diets. Therefore, it is necessary to identify trends across multiple species. Further, urbanisation can differentially affect males and females of the same species [9, 15–17]. Females have a higher reproductive cost than males as they nurse and carry developing offspring until weaning. Thus, there is a sex-based difference in resource requirements, and females tend to prefer areas with better foraging habitat, roosting sites, and water [4, 15–17]. Females need water for successful reproduction and lactation, and many maternity colonies are located near water bodies [4, 10, 15–17]. Subtle differences between males and females in terms of preference for foraging and roosting habitats are now being observed in multiple insectivorous bats, with females in urban landscapes preferring better quality habitat and roosting sites compared to males [4, 10, 15–17].

3.2 Urban Areas as Ecological Traps

Urban habitats may act as ecological traps for synanthropic species and may be detrimental to species in the long run [4]. Reduction in reproductive output in urban spaces can eventually be detrimental to species persistence, and strong conservation actions are needed to prevent further biodiversity loss. Many species may appear to be thriving in urban habitats, but this may not be the case. With the handful of studies comparing the reproductive output of synanthropic bats, varied effects of urbanisation are being observed. In contrast to our *C. sphinx* preliminary data, *M. lucifugus* reproductive success (measured as number of juveniles relative to number of adult females) did not differ between urban and rural areas in Alberta, Canada. Reproductive success was, however, higher in transition zones between urban and rural areas [18]. By contrast, earlier parturition and higher fecundity were observed in *P. kuhlii* populations from urban areas. Urban habitat provided *P. kuhlii* with ample insects to increase reproductive output [19]. More studies are required to quantify the impact of urbanisation on the social and mating systems of bats, including reproductive success, because these can ultimately affect the genetic diversity and persistence of urban populations.

Acknowledgements This study was based on the data collected during the PhD duration of KMG. The authors thank Prof. Uma Ramakrishnan for her support to carry out this research. The study was supported by NCBS-TIFR and DST (SR/S0/AS-65/2021). KMG acknowledges the support of DBT-Ramalingaswami Fellowship (No. BT/HRD/35/02/2006). BC acknowledges the support of Trivedi School of Biosciences, Ashoka University.

We are grateful to the Registrar of IISc for allowing us to work on the bat colony present on the campus. We also thank Dhanabalan, Sivarajan, Pilot Dohvi, Avik Ray, Rajasri Ray, Subhajit Saha, Sandeep Kumar Rana, and M.Sc. Genomics students for their support in the fieldwork.

Literature Cited

1. Ellis EC, Kaplan JO, Fuller DQ, Vavrus S, Goldewijk KK, Verburg PH (2013) Used planet: a global history. Proc Natl Acad Sci U S A 110(20):7978–7985
2. McKinney ML (2002) Urbanization, biodiversity, and conservation the impacts of urbanization on native species are poorly studied, but educating a highly urbanized human population about these impacts can greatly improve species conservation in all ecosystems. Bioscience 52(10):883–890
3. Szulkin M, Munshi-South J, Charmantier A (eds) (2020) Urban evolutionary biology. Oxford University Press, USA
4. Russo D, Ancillotto L (2015) Sensitivity of bats to urbanization: a review. Mamm Biol 80(3):205–212
5. Sepp T, McGraw KJ, Giraudeau M (2020) Urban sexual selection. In: Szulkin M, Munshi-South J, Charmantier A (eds) Urban evolutionary biology. Oxford University Press, USA, pp 234–252
6. Kappeler PM, van Schaik CP (2002) Evolution of primate social systems. Int J Primatol 23(4):707–740
7. Emlen ST, Oring LW (1977) Ecology, sexual selection, and the evolution of mating systems. Science 197(4300):215–223
8. Clutton-Brock TH (1989) Review lecture: mammalian mating systems. Proc Royal Soc B 236(1285):339–372
9. Banks SC, Piggott MP, Stow AJ, Taylor AC (2007) Sex and sociality in a disconnected world: a review of the impacts of habitat fragmentation on animal social interactions. Can J Zool 85(10):1065–1079
10. Jung K, Threlfall CG (2018) Trait-dependent tolerance of bats to urbanization: a global meta-analysis. Proc Royal Soc B 285(1885):20181222
11. Jones G, Jacobs DS, Kunz TH, Willig MR, Racey PA (2009) Carpe noctem: the importance of bats as bioindicators. Endanger Species Res 8(1–2):93–115
12. McCracken GF, Wilkinson GS (2000) Bat mating systems. In: Crichton EG, Krutzsch PH (eds) Reproductive biology of bats. Academic Press, USA, pp 321–362
13. Kerth G (2008) Causes and consequences of sociality in bats. Bioscience 58(8):737–746
14. Kunz TH, Fenton MB (eds) (2005) Bat ecology. University of Chicago Press, USA
15. Lintott PR, Bunnefeld N, Fuentes-Montemayor E, Minderman J, Mayhew RJ, Olley L et al (2014) City life makes females fussy: sex differences in habitat use of temperate bats in urban areas. R Soc Open Sci 1(3):140200
16. Rocha R, Ferreira DF, López-Baucells A, Farneda FZ, Carreiras JM, Palmeirim JM et al (2017) Does sex matter? Gender-specific responses to forest fragmentation in Neotropical bats. Biotropica 49(6):881–890
17. Patriquin KJ, Guy C, Hinds J, Ratcliffe JM (2019) Male and female bats differ in their use of a large urban park. J Urban Ecol 5(1):juz015
18. Coleman JL, Barclay RM (2011) Influence of urbanization on demography of little brown bats (*Myotis lucifugus*) in the prairies of North America. PLoS One 6(5):e20483
19. Ancillotto L, Tomassini A, Russo D (2016) The fancy city life: Kuhl's pipistrelle, *Pipistrellus kuhlii*, benefits from urbanisation. Wildl Res 42(7):598–606
20. Storz JF, Kunz TH (1999) *Cynopterus sphinx*. Mamm Species 613:1–8
21. Balasingh J, Koilraj J, Kunz TH (1995) Tent construction by the short-nosed fruit bat *Cynopterus sphinx* (Chiroptera: Pteropodidae) in Southern India. Ethology 100(3):210–229
22. Garg KM, Chattopadhyay B, Doss DS, Kumar AV, Kandula S, Ramakrishnan U (2012) Promiscuous mating in the harem-roosting fruit bat, *Cynopterus sphinx*. Mol Ecol 21(16):4093–4105
23. Garg KM, Chattopadhyay B, Ramakrishnan U (2018) Social structure of the harem-forming promiscuous fruit bat, *Cynopterus sphinx*, is the harem truly important? R Soc Open Sci 5(2):172024

24. Garg KM, Chattopadhyay B, Doss DS, Kumar AV, Kandula S, Ramakrishnan U (2015) Males and females gain differentially from sociality in a promiscuous fruit bat *Cynopterus sphinx*. PLoS One 10(3):e0122180

25. R Core Team (2013) R: a language and environment for statistical computing. https://www.r-project.org/

26. Storz JF, Balasingh J, Bhat HR, Nathan PT, Doss DPS, Prakash AA et al (2001) Clinal variation in body size and sexual dimorphism in an Indian fruit bat, *Cynopterus sphinx* (Chiroptera: Pteropodidae). Biol J Linn Soc 72(1):17–31

27. Storz JF (2002) Contrasting patterns of divergence in quantitative traits and neutral DNA markers: analysis of clinal variation. Mol Ecol 11(12):2537–2551

28. Chattopadhyay B, Garg KM, Kumar AV, Doss DPS, Rheindt FE, Kandula S et al (2016) Genome-wide data reveal cryptic diversity and genetic introgression in an Oriental cynopterine fruit bat radiation. BMC Evol Biol 16(1):1–15

29. Queller DC, Goodnight KF (1989) Estimating relatedness using genetic markers. Evolution 43(2):258–275

30. Wang J (2011) COANCESTRY: a program for simulating, estimating and analysing relatedness and inbreeding coefficients. Mol Ecol Resour 11(1):141–145

31. Chattopadhyay B, Garg KM, Doss PS, Ramakrishnan U, Kandula S (2011) Molecular genetic perspective of group-living in a polygynous fruit bat, *Cynopterus sphinx*. Mamm Biol 76(3):290–294

32. Dirzo R, Young HS, Galetti M, Ceballos G, Isaac NJ, Collen B (2014) Defaunation in the Anthropocene. Science 345(6195):401–406

Part II
How do Bats Inhabit Urban Environments? Uses of Artificial Roosts, Aerial Habitats, and Green Spaces

Chapter 6
Bat Boxes as Roosting Habitat in Urban Centres: 'Thinking Outside the Box'

Cori L. Lausen, Pia Lentini, Susan Dulc, Leah Rensel, Caragh G. Threlfall, Emily de Freitas, and Mandy Kellner

Abstract Bats in urban environments depend on human-made structures or remnant natural habitats for roosting. Bat boxes are commonly used artificial structures that aim to replace lost tree or building roosts, but they are not a universal solution, or panacea, as few species use them, and other options exist that more closely mimic natural tree cavities. As long-lived mammals, bats may be lured into human-built structures with unstable conditions. These structures could act as 'ecological traps' if they suddenly become inaccessible with few other roost options available. Problems arising from the use of bat boxes, such as mortality events resulting from overheating, may reflect limited roost availability rather than inherent flaws in bat box designs. Mimicking a natural roosting area requires accommodating requisite roost switching. This can be accomplished in urban centres by manipulating existing trees or erecting multiple, varied bat boxes in close proximity, which could require purposeful urban planning. Engaging the public in community-driven bat conservation initiatives may hold the key to ensuring bats thrive in human-dominated

C. L. Lausen (✉)
Wildlife Conservation Society Canada, Kaslo, BC, Canada
e-mail: clausen@wcs.org

P. Lentini
School of Global, Urban, and Social Studies, RMIT University, Melbourne, VIC, Australia

S. Dulc
Thompson Rivers University, Kamloops, BC, Canada

L. Rensel
University of British Columbia Okanagan, Kelowna, BC, Canada

C. G. Threlfall
School of Life and Environmental Science, The University of Sydney, Sydney, NSW, Australia

E. de Freitas
University of Northern British Columbia, Prince George, BC, Canada

M. Kellner
British Columbia Community Bat Program, Revelstoke, BC, Canada

© The Author(s), under exclusive license to Springer Nature Switzerland AG 2022
L. Moretto et al. (eds.), *Urban Bats*, Fascinating Life Sciences,
https://doi.org/10.1007/978-3-031-13173-8_6

landscapes. Here, we discuss problems associated with bat boxes and propose solutions, using case studies from Canada and Australia.

Keywords Bat box · Bat house · Artificial roosts · Overheating · Community-driven conservation

1   Introduction

Habitats for urban hollow- or crevice-roosting bats are in limited supply in cities and towns, where natural features have been replaced by artificial structures such as buildings, bridges, and culverts. Some bats have been able to compensate for this by taking advantage of roosting opportunities in older buildings, and some species are able to exploit the myriad of microclimates that these artificial roosts offer. These building-roosting bats may benefit from lower predation risk and ideal microclimates [1], but the occupancy of building roosts is not without risk. These roosting spaces may be lost to building demolition or renovations, particularly since modern building designs often eliminate entrances to cavities (e.g. Energy Star for homes [2]), and potential alternative roosts may not be available nearby. Bat boxes, sometimes also referred to as bat houses, are an increasingly popular measure used to replace roosting habitat for bats that are evicted from buildings [3] or to compensate for a paucity of natural features. Bat boxes are also used for habitat enhancement, particularly in areas where few tree hollows exist [4].

Bat boxes – rectangular, often wooden containers in which bats can roost (Figs. 6.1 and 6.2) [4, 5] – are designed to provide roosting spaces for bats akin to a hollow or crevice in a large diameter tree, but how effective are they as replacement roosts? Despite recommendations for their installation, few studies have examined how well bats fare in these structures against a range of fitness measures (such as reproductive success), and guidance for their use has often focused on capacity or how many individuals they can accommodate [6]. There may be risks associated with large groups of bats occupying a bat box; for example, overcrowding can reduce the ability of the colony to dissipate heat, and mass mortality from overheating has been observed in various areas of the world (e.g. Spain [7], Australia [8]). Here, we explore the use of bat boxes as replacement roosts. Through case studies from Australia and Canada, two countries where bat box use has raised concerns in relation to a warming climate [8, 9], we demonstrate problems and opportunities that arise when urban bat boxes are used and suggest some potential future directions.

Fig. 6.1 Common North American bat box style – Bat Conservation International's four-chamber maternity box. Two boxes are mounted back to back, with an enclosed space between, to increase microclimate options. (Photo by Jared Hobbs)

2   Bat Roosting Behaviours in Light of Roost Availability

Bat roosting behaviours reflect the type of roosts available [10]. Bats typically express a high degree of site fidelity [11] and natal philopatry [12], returning to the same set of roosts in a home range as long as they remain available and suitable (but see [13]). Reproductive females seek appropriately warm microclimates to expedite gestation and support lactation and growth of pups [14]. To achieve the most suitable microclimates and to reduce predation risk and parasite loads, bats typically switch roosting locations [11, 15]. Roost-switching behaviours often differ between natural and artificial roosts. While colonies using natural roosts may spread out and occupy several roosts in a small area (e.g. [16]), bats using human-built structures will typically use fewer alternate roosts and move less frequently between roosts [17]. For example, reproductive (maternity) colonies in attics will switch roosting locations within the same attic space [e.g. 1], and those in bat boxes may switch roosts every 1–3 days, if such roosts are available, to find optimal microclimates

Fig. 6.2 Two-chamber
rocket box. (Photo by
Susan Dulc)

which are critical for reproductive success and development [13, 18]. Bats in natural
roosts typically move among crevice or cavity roosts (e.g. switch trees) every
1–2 days (e.g. [11, 19, 20]).

The tendency of bats that use artificial roosts to use fewer roosts may strengthen
social associations between bats [21], but dependency on few roost locations also
puts the colony's social structure at risk. For example, Webber et al. [17] compared
social behaviour of big brown bat (*Eptesicus fuscus*) maternity colonies between
tree and building roosts, and bats in buildings formed denser, more highly connected social networks. Over many generations, colonies of building-roosting bats
can become heavily dependent on these structures, roosting in the same roof or attic
for much of the reproductive season and returning to the same structure across consecutive seasons (e.g. [22]). This dependency on a single roosting structure may put
the colony at risk should this structure be lost and alternate roosts be unavailable.
This is in contrast to bats in forests, where the social network of a maternity colony
remains intact until approximately 20% of roosts are removed [23].

Limited availability of artificial roosts may pose additional challenges in light of
seasonally changing roost requirements. The suitability of a roost for a bat will vary
depending on time of year, sex, and reproductive status [24]. Reproductive females
avoid deep or prolonged bouts of torpor because even though it saves energy, torpor

delays offspring development and reduces milk production [25]. Instead, reproductive females typically seek roost microclimates within a particular temperature range to offset the metabolic demands of gestation, lactation, and juvenile development [1, 15, 26, 27]. Within this temperature range, referred to as the thermoneutral zone, the bat is not expending additional energy to generate or dissipate heat. Thermoneutral zones can vary among species, with some desert-adapted bats tolerating high temperatures [e.g. 45.8 °C for free-tailed bats (*Mormopterus* sp.) [28]] that would be lethal to others [e.g. 44.5 °C for little brown myotis (*Myotis lucifugus*), Yuma myotis (*M. yumanensis*), and fringed myotis (*M. thysanodes*) [29]]. Many species employ an additional strategy of cooperatively roosting with conspecifics during the breeding season in maternity colonies to influence microclimates and reduce heat loss and energy expenditure [30]. Lactating females often prefer roosting in buildings to bat boxes [18, 31], possibly due to the insulating nature of building roosts. In comparison, bat boxes more closely track ambient temperatures [26, 31, 32] and typically provide cooler night-time conditions that are less conducive to growth of pups [15, 25]. In contrast, cooler roosts are needed to facilitate the use of torpor for energy savings for males, non-reproductive females, and reproductive females in early pregnancy or post-lactation or in times of food scarcity [25]. If roosts with specific microclimatic conditions are limited or eliminated, reproductive success and survival may suffer.

3   Providing a New Space to Roost: Bat Box Design

Bat boxes have a long history of being used to compensate for the loss of natural roosts or eviction of bats from buildings [32]. Bat boxes are intended to mimic natural hollow or cavity roosts, have been deployed worldwide [4, 5], and vary in size, shape, internal volume, number of chambers, colour, and construction materials. Because of high rates of occupancy when affixed to buildings or poles [5], boxes can be readily deployed in urban centres where suitable roost trees may not exist. Because species differ with respect to roosting requirements, environmental conditions (e.g. solar exposure, daily weather), seasonality (e.g. reproductive stage), and box design (e.g. volume, ventilation, colour, addition of jackets, mounting techniques) all influence the likelihood of occupancy.

The most commonly used boxes have one or more rectangular chambers, often based on the designs of Stebbings and Walsh [33], or, in North America, the four-chamber and vented maternity box design of Bat Conservation International (Fig. 6.1) [32]. These designs provide an interior with enough space for large maternity colonies [32], while boxes with multiple chambers and vertical orientation provide a variety of microclimate options [31, 34]. The number of chambers and internal volume can also be modified to promote certain species assemblages [32].

A second popular option in North America is the rocket box (Fig. 6.2). It features concentric chambers (typically two), allowing bats to access different temperature regimes within the structure. Although some studies report bats preferring rocket boxes [34], others document a preference for four-chamber maternity boxes [35], though more research is needed to identify the attributes that encourage occupancy of rocket boxes and other large, maternity-style boxes [34]. A recent comparison [36] of box modifications determined that adding a water jacket around a rocket box may be beneficial by slightly raising night temperatures, which promotes growth of pups, while slightly lowering daytime temperatures for adult females.

In recent years, large bat condos and mini condos have also gained popularity in North America (Fig. 6.3). These structures aim to replicate the conditions found in

Fig. 6.3 A 'bat condo', an elevated structure built with many chambers inside to house thousands of bats. (Photo by Jared Hobbs)

buildings by including multiple sets of roosting baffles and, in many cases, interior flight space, with additional roosting space under the roof and siding. Bat condos offer a wider variety of microclimates than bat boxes do and can host extremely large maternity colonies [32]. Several bat condos have been occupied by large colonies (thousands) of bats in North America [37] (S. Dulc, unpublished data).

Regardless of design, early guidance from Bat Conservation International suggested using multiple (three or more) bat boxes to effectively replace a lost building roost [32], but this guidance has changed. Today, wildlife/bat organisations tend to recommend installing one multi-chamber box, likely as a 'quick fix' that entices landowners to replace roosts lost through evictions of building-roosting bats. To simulate the myriad of roosting options in a natural 'roosting area' [16] or within an attic roost [1], it is necessary to promote the use of multiple boxes of varying styles and solar exposure in close proximity. This 'Goldilocks' approach may provide the roost temperatures that are 'just right' and promote successful gestation, lactation, pup rearing, and preparation for hibernation. In addition, providing multiple bat boxes could also facilitate beneficial behaviours such as roost switching, which reduces predation risk and parasite loads.

4   Bat Boxes Are Not a Universal Solution or 'Panacea'

Here, we describe the two key motivations for the installation of bat boxes in urban areas: (1) to replace roosting habitat for bats that are evicted from buildings and (2) to provide supplementary roosts for local bat assemblages in areas where natural features have been lost or are lacking. We also elaborate on concerns that surround 'quick fix' adoption of bat boxes in urbanising landscapes.

4.1 Successfully Replacing an Eliminated Roost Post-eviction

Findings from research on bat boxes as effective replacement roosts are equivocal – eliminated roosts are unlikely to be adequately replaced by one bat box [38]. If boxes do not provide the same microclimates used in the building roost from which bats were evicted, this inadequate roost replacement could be detrimental to bats. This is especially true where the eliminated roost(s) provided ideal conditions for maternity colonies and few alternate roosts are available [39]. If replacement boxes are not high-quality habitat and more suitable alternate roosts are nearby, boxes may remain unoccupied. Only 1% [40] to 46% [13] of marked, evicted individuals are recaptured near an erected replacement roost. Although few studies have assessed the impacts of eviction on reproductive success, evidence suggests that reproductive rate declines post-eviction when bats are forced to use alternate roosts ([41], C. Lausen, unpublished data).

We further suggest that landowners reconsider the need to exclude bats from buildings because there may be alternative solutions that would allow bats to remain safely in existing urban roosts. Indeed, some landowners have lived successfully with bats in their buildings for years without negative consequence, by employing relatively simple adaptations, such as ensuring separation of human and bat living spaces and yearly cleaning of guano after bats have left for the season (BC Community Bat, M. Kellner, pers. obs.; also [42]). Wherever it is practical, existing roost sites in the urban landscape should be identified and protected, with bat boxes as a complementary measure. If exclusion is indeed necessary and bat boxes are used, then minimising the risk of boxes becoming ecological traps hinges on ensuring an adequate number and diversity of boxes and placements and installing them at a height inaccessible to ground-based predators (i.e. ideally ≥5 m from the ground; predator-guard at base of pole, or away from roof if building-mounted).

4.2   How a Bat Box Might Become an 'Ecological Trap'

How can we know whether a bat box is an ecological trap? Ecological traps occur when animals select poor-quality habitat (or other resources) over higher-quality resources [43]. Occupancy of bat boxes is often naïvely equated with success and may be incorrectly interpreted as a preference for this roost type. Bats may choose to roost in a box for reasons other than preference, such as site fidelity after exclusion from an adjacent building, lack of alternate roosts within a preferred foraging area, or social cohesion. The cues that drive bats to use bat boxes are not well understood and may not accurately reflect the quality of the box as roosting habitat. Fidelity of bats to a roosting area [11] may lead to maladaptive roost choice when high-quality roosts are lost [41]. Bats may subsequently experience reduced reproductive fitness or increased mortality risk, perhaps due to poor box placement, insufficient numbers of boxes, and/or low variation among box microclimates.

Bat boxes may function as ecological traps if they increase predation risk. For example, bats in boxes may be at increased risk of predation by urban animals, such as corvids, owls [44], and rodents [45], that are attracted to occupied boxes by scent or sound, and domestic cats (*Felis catus*) which can reach particularly high densities in urban areas [46]. Free-ranging domestic cats are opportunistic, subsidised predators that can have a significant impact on maternity colonies of bats [47, 48]. In Sydney, Australia, scent-hunting predators, including possums, rodents, and corvids, were attracted to bat boxes containing guano – interestingly, they were more likely to investigate boxes with small rather than large accumulations of guano, suggesting a negative correlation between group size and vulnerability to predation [45]. Although no studies have explicitly assessed cat predation on bat boxes, a recent review identified 86 bat species, including 19 (40%) of Australia's hollow-roosting bats [46], that are known to be preyed upon or threatened by cats globally.

The continued use of bat boxes as habitat replacement or enhancement tools with incomplete knowledge of bat roosting and environmental context may unintentionally facilitate the creation of ecological traps. In other words, it is only by paying attention to the behaviour and health of a colony of bats in artificial structures that we can understand where bats truly thrive.

4.3 Supplementing Existing Habitat: Environmental Context, Bat Assemblages, and Box Design

Environmental context, bat assemblage, and box designs and arrangements are critical considerations when supplementing existing roosts in an area to promote bat abundance. Landscape context and box design can lead to the preferential use and domination of bat boxes by only one or a few species [49]. This has triggered concerns that the installation of boxes may cause an overall shift in community composition. For example, while some tree crevice-roosting bats in the United States and Canada have adapted to human-built structures, only 15 of 47 species (32%) have been documented using bat boxes (M. Kellner, unpublished data; also [32]). This concern was also raised in the Southern Hemisphere [50, 51]. However, evidence from Australian studies suggest that this shift may not occur. One 5-year study on 18 sites across Greater Melbourne found no evidence of assemblage shift [52]. Similarly, in a comparison between reserves with and without bat boxes in Sydney, reserves with boxes exhibited no shift in assemblage composition, but they did have greater activity of one species, Gould's wattled bat (*Chalinolobus gouldii*), that occupies boxes and less activity of other bats [53]. This suggests that although species diversity persists in reserves with bat boxes, box availability may affect relative abundance.

4.4 How Hot Is Too Hot?

Bat boxes with full solar exposure throughout the day maintain warm temperatures. These conditions may be advantageous during some parts of the reproductive cycle, but there is a fine line between an 'optimally warm microclimate' and too hot. Reproductive bats select roosts based on microclimate [15], and even natural roosts can reach extremely high temperatures, which are likely to be lethal for most species (e.g. rock crevice roosts >50 °C [54] but see [28]). Having a selection of available roosts with varying microclimates typically ensures that bats can switch roosts to appropriately thermoregulate and avoid overheating. However, global warming may raise the risk of overheating for bats in boxes [7, 8, 55] (see Box 6.1: Case Study – Overheating Bat Boxes). If roost choice becomes limited, as is increasingly true in many cities, maternity colonies may remain in bat boxes that overheat, leading to heat stress and potential mortality.

Box 6.1: Case Study – Overheating Bat Boxes

Overheating of occupied bat boxes seems to be a new phenomenon in southern British Columbia (BC), Canada, with record summer heat in recent years. In Greater Vancouver, a mixed *Myotis lucifugus* and *M. yumanensis* colony (~600 individuals total, L. Rensel, unpublished data) used a set of three identical, south-facing, multi-chambered bat boxes that replaced a lost adjacent building roost in an area of relatively few buildings and trees. Similarly, S. Dulc (unpublished data), working in Creston, monitored a *M. yumanensis* colony (~300 individuals) occupying one multi-chambered bat box erected near a building from which the bats were evicted. In both cases, the bat colonies occupied boxes continuously, raising young throughout the summer, suggesting that alternate roosts were not available or not being used.

In 2018 and 2019, bat box microclimates were recorded, and overheating was observed in both Creston and Vancouver maternity colonies. In 2019, in Creston, pregnant females with wet fur (from urinating on themselves to facilitate evaporative cooling (e.g. [54])) were observed crowding at the box entrance midday, in mid-June (pre-parturition). Several bats then flew to roost on shaded bark of adjacent trees as the bat box exceeded 40 °C and 100% humidity (maximum ambient temperature and relative humidity were 31.7 °C and 75%, respectively). During a 2018 summer heatwave in Greater Vancouver, adults and juveniles fell out of the three bat boxes, seemingly due to heat exhaustion (bat box temperature reached 46 °C), and ~75 bats died (J. Saremba, Burke Mountain Naturalists, pers. comm.). This mass mortality event suggested that the bat boxes became too hot for bats.

Immediately after the mass mortality in Vancouver, the boxes were painted a light colour (to reflect light). Although this change successfully reduced box temperatures in times of extreme heat, it also lowered internal box temperatures in spring and, thus, could prolong gestation. Therefore, alternative cooling measures for boxes were identified; the above-mentioned box colour decision was reversed, and, instead, an additional bank of north-facing boxes was built next to the south-facing boxes so that bats can choose from a cooler set of boxes.

In June 2021, when BC citizens were forewarned of another heatwave, volunteers mobilised to shade bat boxes with awnings or white sheets. No mass mortalities of bats were reported during this heatwave. In some cases, this could be attributed to the precautionary actions. For example, at one site where microclimate was being recorded, the temperature in the bat box dropped from 44.1 °C (nearing the lethal 44.5 °C [24]) to 42.6 °C within 30 min of installing a sunshield (S. Dulc, unpublished data), although adults and young were still observed crowding the box entrance and vent (Fig. 6.4). In other cases, where alternate roosts presumably existed, the bats simply abandoned their bat boxes.

(continued)

Box 6.1 (continued)

Fig. 6.4 (**a**) Adult and juvenile bats crowding the opening and vent of an overheating four-chamber maternity bat box in Creston, British Columbia. Fur wetted by urine is a cooling mechanism. (**b**) A temporary white shade erected on the box during this heatwave prevented lethal microclimates. (Photo by Susan Dulc)

(continued)

Box 6.1 (continued)

These Canadian examples of overheating bat boxes and heat stress during unprecedented heatwaves raise the question of what role context plays in determining whether a bat box is detrimental to a maternity colony. Perhaps as there become less roost options available to a colony raising young, the risk of bats roosting in unsuitable microclimates rises. These risks include cold temperatures that do not support growth of young, hot temperatures that exceed thermal tolerances [56], or a lethal combination of temperature and humidity (i.e. heat stress, temperature-humidity index) in which evaporative water loss (and consequently heat loss) no longer occur in high humidity.

It is not yet understood why bats would remain in a bat box that overheats. New local weather patterns and unusual heatwaves that are not anticipated by bats may be partially to blame [55], but this behaviour may also stem from the reduced choice of roosting habitat, with options likely to become increasingly scarce as our landscapes continue to urbanise or be otherwise transformed.

5 Thinking Outside the Box: Recommendations for the Installation and Use of Bat Boxes and Other Complementary Roost Structures

Bat boxes may not be a panacea, as described above, but in urbanising landscapes, they may still play a crucial role in bat conservation. Increasingly, humans control the availability of habitat for bats, and where bats control arthropod pests [32], retaining bat populations is desirable and can be promoted (see Box 6.2: Case Study – Engaging Citizens). Purposeful planning of roosting habitat for bats is necessary (see Sect. 6 below: Urban Planning for Bats). Problems with bat boxes, particularly where they act as ecological traps, may result from placement rather than an inherent design flaw.

The literature presents a variety of reports on the use and success of bat boxes, including design and placement performance relative to geographic location. Species-specific requirements, as well as local climate, weather patterns, landforms, and solar exposure, provide important context for boxes being used by bats. Many studies report high usage of boxes by common or widespread species, especially those tolerant of urbanization [47, 57]. The inherent variability of occupied artificial roosts should not be surprising given that natural roosts and bats are also diverse.

Box 6.2: Case Study – Engaging Citizens

Retaining bats and roosting habitat in urban centres can be further supported through the education and engagement of local citizens. Two successful community bat programmes in western Canada (albertabats.ca in Alberta; bcbats.ca in BC) promote the conservation of urban bats and the creation of bat-friendly communities by providing guidance and support to landowners who live with or wish to evict bats without harming them. As a result, interest in bat conservation and volunteer participation has grown, enabling a broad-scale and long-term summer roost monitoring program (e.g. 360 roosts in BC in 2020). A bat box registration and monitoring programme is now providing data on box occupancy, timing of seasonal arrivals/departures and parturition, and disease surveillance.

In Australia, where 46 bat species (or 57%) partially or wholly rely on tree hollows for roosting and breeding [70], a bat monitoring program at Melbourne's Organ Pipes National Park is one of the world's longest-running bat box programmes [71], and it exemplifies how enhancing bat habitat can raise awareness and appreciation for bats. Since 1992, more than 42 bat boxes have been installed, monitored, and maintained at this park to compensate for the lack of natural hollows. Volunteers have been coordinating bat box inspections and bat assessments since the programme started. Eight species have been recorded using the boxes and >2000 individuals have been marked. The research and data have provided important insights into bat ecology and conservation, including studies of reproduction, including documentation of double-breeding events, survival rates and life history, box occupancy patterns and design considerations, community dynamics, thermal physiology, and tag retention rates. Most importantly, this bat box programme has provided people with the rare opportunity to see bats up close, fostering an appreciation for these creatures.

5.1 What to Recommend for the Installation and Proposed Use of Bat Boxes

So, what universal guidance might be derived for bat boxes? Determining what roost choices already exist in an area and how to supplement or provide variety is likely far more important than the design of one box, although modifications that support passive heating and insulating designs have been proposed and could prove beneficial in some areas (e.g. [58]). Although there are many publications on styles, stains, and other box deployment options (e.g. [55]), these may be species- and location-specific, and do not address the real issue of context. It might be that using many different designs within a large variety of solar exposures may be more strategic than trying to invent or identify *one* design and appropriately place a *single* bat box to achieve the range of microclimates needed by a bat population (e.g. a maternity colony throughout the reproductive season). However, we recognise that cost may limit what can be done without collaboration.

Ultimately, a 'one box does all' goal is neither achievable nor desirable because it is likely to appeal to a small subset of species, have narrow temporal use, and possibly reduce bat fitness. For example, a box stained with a light colour and positioned to avoid overheating in midsummer will likely be unsuitable in early summer during gestation [59]. Similarly, an insulated box will retain heat for pups at night midsummer but is unlikely to warm up enough from solar incidence to promote gestation. Lausen and Barclay [15] found that a maternity colony of approximately 40 big brown bats (*Eptesicus fuscus*) used more than 70 rock crevice roosts over the course of the summer. In a forested area of northern Alberta, Canada, Olson and Barclay [20] identified 135 roost trees used by two colonies of *Myotis lucifugus* (approximately 400 bats). The documented pattern of boxes being dominated by one or a few species may reflect not so much the narrow range of species that will use boxes (e.g. [60]), as the narrow range of box designs and placements usually deployed.

Small structural differences in artificial habitat (microhabitat) can dictate suitability of a box as a roost; for example, culverts used by the fishing bat Gould's large-footed myotis (*Myotis macropus*) in Australia differed from unoccupied culverts in the size of holes in the concrete and the presence of crevices [61]. This highlights the need to consider microhabitat and macrohabitat properties when planning bat habitat in an urban environment.

It is also important to consider the spatial placement of bat boxes. The distance between roosts can matter if bats are forced to move young or switch roosts during daylight when predation risk could be high. Typical distances between successive natural day roosts have been reported for some North American bats; e.g. <200 m for pallid bats (*Antrozous pallidus*) [62], <110 m for western small-footed myotis (*Myotis ciliolabrum*), ~1.25 km for *Eptesicus fuscus* [63], and ~90 m for long-eared myotis (*Myotis evotis*) [19]. In natural situations, crevices may be spatially close so that, if needed, a daytime roost switch would be possible. Successful use of bat boxes in urban centres may require that the proximity of bat boxes mimics that of roosts in natural areas.

5.2 Beyond the Box: Installing Other Artificial Roost Structures to Diversify Available Roost Space

Artificial structures besides bat boxes may also complement existing roosting opportunities and promote roosting by a greater diversity of bats. Wooden shingles attached to the outside of buildings as siding/roofing may offer roosting habitat and attract bark-roosting species, such as the silver-haired bat (*Lasionycteris noctiva-gans*) and long-legged myotis (*Myotis volans*; M. Kellner, pers. obs.), that rarely use bat boxes. Bark mimic structures, made of resin to resemble loose bark, can be affixed to trees [64]. The bark mimic, BrandenBark™, used in the eastern United States for the Indiana myotis (*Myotis sodalis*), has hosted other crevice-roosting bat species [65]. These structures can be installed on tall poles, in urban environments, simulating trees where there are none. Where trees exist, roosting habitat can also be enhanced through modifications that create roosting locations (cavities/crevices), which may be possible in some urban green spaces. In Australia, mechanically created hollows (chainsaw cuts) in tree trunks can mimic natural cavities used by long-eared bats (*Nyctophilus* species) [66]. By structurally and thermally resembling natural roosts, these hollows may benefit bats [67]. In Costa Rica, sawdust concrete bat roosts resembling tree hollows have been shown to effectively host frugivorous and nectivorous bats [68]. Few data exist on the effectiveness of artificial roosts for foliage-roosting species, but the Ussuri tube-nosed bat (*Murina ussuriensis*) used roosts made to mimic the leaves of butterbur (*Petasites japonicus*) and Japanese whitebark magnolia (*Magnolia obovata*) [69]. Although foliage roosts are considered abundant [11], they may not be in urban environments, so these roosts may be beneficial here.

6   Conclusion: Urban Planning for Bats

As landscapes continue to urbanise or otherwise change and natural habitats become increasingly limited, availability of roosts for bats is likely to diminish – hence the need to plan bat box designs and installations to effectively support urban bats. Evaluating existing foraging and roosting habitat in the context of a building eviction is necessary to appropriately mitigate for eliminated roosts. A bat box erected for reasons other than to mitigate an eviction may never become occupied, because a bat box is but *one* crevice roost, akin to one tree cavity. As shown by Olson and Barclay [20], it takes a forest of many tree roosts to meet the needs of one bat colony. We recognise that bat boxes are not insignificant in cost, and can be challenging to install and maintain; while they are not the only option for artificial roosts, variable designs and placements of boxes make them an attractive option for replacing lost roosting habitat. Just as our city planning is now occurring in some places to ensure buildings are safe and near resources that people require, urban landscapes should also be planned for wildlife, including bats, to ensure the availability of safe and accessible roosts near suitable food and water resources. There can be immense value in engaging stewardship groups and other volunteers in planning and implementing urban bat conservation programmes.

Literature Cited

1. Lausen CL, Barclay RMR (2006) Benefits of living in a building: big brown bats (*Eptesicus fuscus*) in rocks versus buildings. J Mammal 87(2):362–370. https://doi.org/10.1644/05-MAMM-A-127R1.1

2. U.S. Department of Energy (2013) Energy efficiency and renewable energy. Thermal bypass air barriers in the 2009 International Energy Conservation Code. Building America Solutions Centre, DOE Building Technologies Program PNNL-SA_90573. Available: https://www.energy.gov/sites/prod/files/2014/01/f6/4_3d_ba_innov_thermalbypassairbarriers_011713.pdf

3. Arias M, Gignoux-Wolfsohn S, Kerwin K, Maslo B (2020) Use of artificial roost boxes installed as alternative habitat for bats evicted from buildings. Northeast Nat 27(2):201–214. https://doi.org/10.1656/045.027.0203

4. Rueegger N (2016) Bat boxes—a review of their use and application, past, present and future. Acta Chiropt 18(1):279–299. https://doi.org/10.3161/15081109ACC2016.18.1.017

5. Mering ED, Chambers CL (2014) Thinking outside the box: A review of artificial roosts for bats. Wildl Soc Bull 38(4):741-751. https://doi.org/10.1002/wsb.461

6. Tuttle MD (1988) America's neighborhood bats. University of Texas Press, Austin, TX, USA

7. Flaquer C, Puig X, López-Baucells A, Torre I, Freixas L, Mas M, Porres X, Arrizabalaga A (2014) Could overheating turn bat boxes into death traps? Barb 7(1). https://doi.org/10.14709/BarbJ.7.1.2014.08

8. Griffiths SR, Rhodes M, Parsons S (2021) Overheating turns a bat box into a death trap. Pac Conserv Biol. Published online 23 March 2021. https://doi.org/10.1071/PC20083

9. Lausen CL, Nagorsen DN, Brigham RM, Hobbs J (2022) Bats of British Columbia, 2nd edn. Royal BC Museum, Victoria

10. Kunz TH, Lumsden LF, Fenton MB (2006) Ecology of cavity and foliage roosting bats. In: Kunz TH, Fenton MB (eds) Bat ecology. University of Chicago Press, Chicago, IL, pp 3–89

11. Lewis SE (1995) Roost fidelity of bats: a review. J Mammal 76(2):481–496. https://doi.org/10.2307/1382357

12. Burland TM, Wilmer W (2001) Seeing in the dark: molecular approaches to the study of bat populations. Biol Rev Camb Philos Soc 76(3):389–409. https://doi.org/10.1017/s1464793101005747

13. Slough BG, Jung TS (2020) Little brown bats utilize multiple maternity roosts within foraging areas: implications for identifying summer habitat. J Fish Wildl Manag 11(1):311–320. https://doi.org/10.3996/052019-JFWM-039

14. Zahn A (1999) Reproductive success, colony size and roost temperature in attic-dwelling bat *Myotis myotis*. J Zool 247(2):275–280. https://doi.org/10.1111/j.1469-7998.1999.tb00991.x

15. Lausen CL, Barclay RMR (2003) Thermoregulation and roost selection by reproductive female big brown bats (*Eptesicus fuscus*) roosting in rock crevices. J Zool 260(3):235–244. https://doi.org/10.1017/S0952836903003686

16. Willis CKR, Brigham RM (2004) Roost switching, roost sharing and social cohesion: forest-dwelling big brown bats, *Eptesicus fuscus*, conform to the fission–fusion model. Anim Behav 68(3):495–505. https://doi.org/10.1016/j.anbehav.2003.08.028

17. Webber QMR, Brigham RM, Park AD, Gillam EH, O'Shea TJ, Willis CKR (2016) Social network characteristics and predicted pathogen transmission in summer colonies of female big brown bats (*Eptesicus fuscus*). Behav Ecol Sociobiol 70(5):701–712. https://doi.org/10.1007/s00265-016-2093-3

18. Rensel L (2021) Roost selection and social organization of myotis in maternity colonies. Master's thesis. University of British Columbia Okanagan, Kelowna, British Columbia, Canada

19. Nixon AE, Gruver JC, Barclay RMR (2009) Spatial and temporal patterns of roost use by western long-eared bats (*Myotis evotis*). Am Midl Nat 162(1):139–147. https://doi.org/10.1674/0003-0031-162.1.139

20. Olson CR, Barclay RMR (2013) Concurrent changes in group size and roost use by reproductive female little brown bats (*Myotis lucifugus*). Can J Zool 91(3):149–155. https://doi.
   org/10.1139/cjz-2012-0267
21. Kerth G, Perony N, Schweitzer F (2011) Bats are able to maintain long-term social relationships
   despite the high fission–fusion dynamics of their groups. Proc R Soc B 278(1719):2761–2767.
   https://doi.org/10.1098/rspb.2010.2718
22. Schorr RA, Siemers JL (2021) Population dynamics of little brown bats (*Myotis lucifugus*) at
   summer roosts: apparent survival, fidelity, abundance, and the influence of winter conditions.
   Ecol Evol 11(12):7427–7438. https://doi.org/10.1002/ece3.7573
23. Bondo KJ, Willis CKR, Metheny JD, Kilgour RJ, Gillam EH, Kalcounis-Rueppell MC,
   Brigham RM (2019) Bats relocate maternity colony after the natural loss of roost trees. J Wild
   Manag 83(8):1753–1761. https://doi.org/10.1002/jwmg.21751
24. Lausen CL, Barclay RMR (2002) Roosting behaviour and roost selection of female big
   brown bats (*Eptesicus fuscus*) roosting in rock crevices in southeastern Alberta. Can J Zool
   80(6):1069–1076. https://doi.org/10.1139/z02-086
25. Racey PA (1982) Ecology of bat reproduction. In: Kunz TH (ed) Ecology of bats. Boston,
   Springer, pp 57–104. https://doi.org/10.1007/978-1-4613-3421-7
26. Lourenço SI, Palmeirim JM (2004) Influence of temperature in roost selection by *Pipistrellus
   pygmaeus* (Chiroptera): relevance for the design of bat boxes. Biol Conserv 119(2):237–243.
   https://doi.org/10.1016/j.biocon.2003.11.006
27. Pretzlaff I, Kerth G, Dausmann KH (2010) Communally breeding bats use physiological
   and behavioural adjustments to optimise daily energy expenditure. Naturwissenschaften
   97(4):353–363. https://doi.org/10.1007/s00114-010-0647-1
28. Bondarenco A, Körtner G, Geiser F (2014) Hot bats: extreme thermal tolerance in a desert
   heat wave. Naturwissenschaften 101(8):679–685. https://doi.org/10.1007/s00114-014-1202-2
29. O'Farrell MJ, Studier EH (1970) Fall metabolism in relation to ambient temperatures in three species of Myotis. Comp Biochem Physiol 35(3):697–703. https://doi.
   org/10.1016/0010-406X(70)90987-4
30. Willis CKR, Brigham RM (2007) Social thermoregulation exerts more influence than microclimate on forest roost preferences by a cavity-dwelling bat. Behav Ecol Sociobiol 62(1):97–108.
   https://doi.org/10.1007/s00265-007-0442-y
31. Brittingham MC, Williams LM (2000) Bat boxes as alternative roosts for displaced bat maternity colonies. Wildl Soc Bull 28(1):197–207. http://www.jstor.org/stable/4617303
32. Tuttle MD, Kiser S, Kiser S (2013) The bat house builder's handbook, 3rd edn. Bat Conservation
   International, Austin. Available: https://batweek.org/wp-content/uploads/2018/01/
   BHBuildersHdbk13_Online.pdf
33. Stebbings B, Walsh S (1985) Bat boxes: a guide to their history, function, construction and use
   in the conservation of bats. Flora and Fauna Preservation Society, London
34. Hoeh JPS, Bakken GS, Mitchell WA, O'Keefe JM (2018) In artificial roost comparison, bats
   show preference for rocket box style. PLoS One 13(10):e0205701. https://doi.org/10.1371/
   journal.pone.0205701
35. Weier SM, Linden VMG, Grass I, Tscharntke T, Taylor PJ (2019) The use of bat houses as day
   roosts in macadamia orchards. S Afr PeerJ 7:e6954. https://doi.org/10.7717/peerj.6954
36. Tillman FE, Bakken G, O'Keefe, JM (2021) Design modifications affect bat box temperatures
   and suitability as maternity habitat. Ecol Solutions Evid 2(4):p.e12112
37. Pennisi L, Holland S, Stein T (2004) Achieving bat conservation through tourism. J Ecotour
   3(3):195–207
38. Griffiths SR, Bender R, Godinho LN, Lentini PE, Lumsden LF, Robert KA (2017) Bat boxes
   are not a silver bullet conservation tool. Mammal Rev 47(4):261–265. https://doi.org/10.1111/
   mam.12097
39. Johnson JS, Treanor JJ, Slusher AC, Lacki MJ (2019) Buildings provide vital habitat for little
   brown myotis (*Myotis lucifugus*) in a high-elevation landscape. Ecosphere 10(11). https://doi.
   org/10.1002/ecs2.2925

40. Neilson AL, Fenton MB (1994) Responses of little brown myotis to exclusion and to bat houses. Wildl Soc Bull 22(1):8–14. http://www.jstor.org/stable/3783215

41. Brigham RM, Fenton MB (1986) The influence of roost closure on the roosting and foraging behaviour of *Eptesicus fuscus* (Chiroptera: Vespertilionidae). Can J Zool 64(5):1128–1133. https://doi.org/10.1139/z86-169

42. Bat Conservation Trust (2015) Living with bats – a guide for roost owners. London, UK. Available: https://cdn.bats.org.uk/uploads/pdf/Living-with-Bats.pdf?v=1541085207

43. Robertson BA, Hutto RL (2006) A framework for understanding ecological traps and an evaluation of existing evidence. Ecology 87(5):1075–1085. doi:https://doi.org/10.1890/0012-965 8(2006)87[1075:AFFUET]2.0.CO;2

44. Speakman JR (1991) The impact of predation by birds on bat populations in the British Isles. Mammal Rev 21(3):123–142. https://doi.org/10.1111/j.1365-2907.1991.tb00114.x

45. Threlfall C, Law B, Banks PB (2013) Odour cues influence predation risk at artificial bat roosts in urban bushland. Biol Lett 9(3):20121144. https://doi.org/10.1098/rsbl.2012.1144

46. Oedin M, Brescia F, Millon A, Murphy BP, Palmas P, Woinarski JCZ, Vidal E (2021) Cats *Felis catus* as a threat to bats worldwide: a review of the evidence. Mammal Rev 51(3). Published online February 15, 2021:mam.12240. https://doi.org/10.1111/mam.12240

47. Russo D, Ancillotto L (2015) Sensitivity of bats to urbanization: a review. Mamm Biol 80(3):205–212. https://doi.org/10.1016/j.mambio.2014.10.003

48. Welch JN, Leppanen C (2017) The threat of invasive species to bats: a review. Mammal Rev 47(4):277–290. https://doi.org/10.1111/mam.12099

49. Rueegger N, Goldingay R, Law B, Gonsalves L (2020) Testing multi-chambered bat box designs in a habitat-offset area in eastern Australia: influence of material, colour, size and box host. Pac Conserv Biol 26(1):13–21. https://doi.org/10.1071/PC18092

50. Griffiths SR, Lumsden LF, Bender R, Irvine R, Godinho LN, Visintin C, Eastick DL, Robert KA, Lentini PE (2019) Long-term monitoring suggests bat boxes may alter local bat community structure. Aust Mammal 41(2):273. https://doi.org/10.1071/AM18026

51. Law B, Eby P, Lunney D, Lumsden L (2011) Biology and conservation of Australasian bats. Royal Zoological Society of New South Wales, Mosman, pp 288–296 and 424–442

52. Griffiths SR, Lumsden LF, Robert KA, Lentini PE (2020) Nest boxes do not cause a shift in bat community composition in an urbanised landscape. Sci Rep 10(1):6210. https://doi.org/10.1038/s41598-020-63003-w

53. Velasco S (2018) Bat box bluff? An investigation into the facilitated dominance of Gould's wattled bat. Honours thesis. The University of Wollongong, Wollongong, Australia

54. Lausen CL (2001) Thermoregulation and roost selection by reproductive big brown bats (*Eptesicus fuscus*) roosting in the South Saskatchewan River Valley, Alberta: rock-roosting and building-roosting colonies. MSc thesis. University of Calgary, Calgary, AB, Canada

55. Bideguren GM, López-Baucells A, Puig-Montserrat X, Mas M, Porres X, Flaquer C (2019) Bat boxes and climate change: testing the risk of over-heating in the Mediterranean region. Biodivers Conserv 28(1):21–35. https://doi.org/10.1007/s10531-018-1634-7

56. Crawford RD, O'Keefe C (2019) Bat boxes and climate change: testing the risk of over-heating in the Mediterranean region. Biodivers Conserv 28(1):21–35. https://doi.org/10.1007/s10531-018-1634-7

57. Jung K, Threlfall CG (2018) Trait-dependent tolerance of bats to urbanization: a global meta-analysis. Proc R Soc B 285(1885):20181222. https://doi.org/10.1098/rspb.2018.1222

58. Fontaine A, Simard A, Dubois B, Dutel J, Elliott KH (2021) Using mounting, orientation, and design to improve bat box thermodynamics in a northern temperate environment. Sci Rep 11(1):7728. https://doi.org/10.1038/s41598-021-87327-3

59. Griffiths SR, Rowland JA, Briscoe NJ, Lentini PE, Handasyde KA, Lumsden LF, Robert KA (2017) Surface reflectance drives nest box temperature profiles and thermal suitability for target wildlife. PLoS One 12(5):e0176951. https://doi.org/10.1371/journal.pone.0176951

60. Baranauskas K (2007) Bats (Chiroptera) found in bat boxes in southeastern Lithuania. Ekologija 53(4):34–37. https://doi.org/10.2478/v10043-010-0005-8

61. Gorecki V, Rhodes M, Parsons S (2019) Roost selection in concrete culverts by the large-footed myotis (*Myotis macropus*) is limited by the availability of microhabitat. Aust J Zool 67(6):281. https://doi.org/10.1071/ZO20033

62. Lewis SE (1996) Low roost-site fidelity in pallid bats: associated factors and effect on group stability. Behav Ecol and Sociobiol 39(5):335–344. https://doi.org/10.1007/s002650050298

63. Lausen CL (2007) Roosting ecology and landscape genetics of prairie bats. PhD dissertation. University of Calgary, Calgary, AB, Canada

64. Mering ED, Chambers CL (2012) Artificial roosts for tree-roosting bats in northern Arizona. Wildl Soc Bull 36(4):765–772. https://doi.org/10.1002/wsb.214

65. Adams J, Roby P, Sewell P, Schwierjohann J, Gumbert M, Brandenburg M (2015) Success of Brandenbark™, an artificial roost structure designed for use by Indiana bats (*Myotis sodalis*). JASMR 4(1):1–15. https://doi.org/10.21000/JASMR15010001

66. Rueegger N (2017) Artificial tree hollow creation for cavity-using wildlife – trialing an alternative method to that of nest boxes. For Ecol Manag 405:404–412. https://doi.org/10.1016/j.foreco.2017.09.062

67. Griffiths S, Lentini P, Semmens K, Watson S, Lumsden L, Robert K (2018) Chainsaw-carved cavities better mimic the thermal properties of natural tree hollows than nest boxes and log hollows. Forests 9(5):235. https://doi.org/10.3390/f9050235

68. Kelm DH, Wiesner KR, von Helversen O (2008) Effects of artificial roosts for frugivorous bats on seed dispersal in a neotropical forest pasture mosaic. Conserv Biol 22(3):733–741. https://doi.org/10.1111/j.1523-1739.2008.00925.x

69. Matsuoka S (2008) Use of artificial roosts by Ussuri tube-nosed bats *Murina ussuriensis*. Bull For For Prod Res Jpn 7(1):9–12. Available: https://www.ffpri.affrc.go.jp/pubs/bulletin/401/documents/406-2.pdf

70. Churchill SK (2009) Australian bats, 2nd edn. Allen and Unwin, Crows Nest, 255pp. http://www.loc.gov/catdir/toc/fy0905/2009286816.html

71. Bender R (2011) Bat roost boxes at Organ Pipes National Park, Victoria: seasonal and annual usage patterns. In: Law BS, Eby P, Lunney D, Lumsden L (eds) Biology and conservation of Australasian bats. Royal Zoological Society of New South Wales, Mosman, pp 443–459

Chapter 7
Aerial Habitats for Urban Bats

Lauren A. Hooton, Lauren Moretto, and Christina M. Davy

Abstract Aerial habitats in cities are understudied but essential for urban, flying wildlife. Understanding interactions between aerial habitats and wildlife (i.e. aeroecology, the study of aerial ecosystems) can identify key foraging and migratory resources for these species. In an urban context, where urban infrastructures dominate the airspace, an aeroecology lens can inform wildlife-friendly urban planning and land management. In this chapter, we highlight key knowledge gaps associated with urban aeroecology of bats by exploring how elements of urban environments may influence the aerial habitat of bats. We draw on studies from within and outside cities to consider how bats might navigate characteristics of urban areas that alter the airspace, including roads and traffic, anthropogenic noise, artificial light, urban heat islands, air pollution, urban canyon effects, high wind speeds, and windows. Finally, we summarise potential approaches to mitigate the negative impacts of these challenges and support aeroconservation of urban bats and other flying wildlife.

Keywords Aeroecology · Aeroconservation · Anthropogenic urban infrastructure · Urban airspace

L. A. Hooton
Department of Environmental and Life Sciences, Trent University, Peterborough, ON, Canada

L. Moretto
Vaughan, ON, Canada

C. M. Davy (✉)
Department of Biology, Carleton University, Ottawa, Ontario, Canada
e-mail: Christina.Davy@carleton.ca

1   Introduction

The study of aerial ecosystems (aeroecology; [1–3]) and their conservation (aero-conservation [4–7]) can inform our understanding of the behaviour and habitat requirements of flying wildlife, including bats. Bats rely on aerial habitats to access a range of resources. For example, nectivorous and frugivorous bats commute through the airspace to access flowering and fruiting trees. Insectivorous bats aerial-hawk flying prey, or glean prey off aerial-adjacent surfaces, and some species even mate in the air [8].

Structural and environmental modifications in urban basoaerial habitats (0–1 km altitude) [4–6] pose challenges for navigation beyond those faced by bats in non-urban habitats. Each species of bat may perceive these challenges differently, and flight style may determine how much of the airspace that a species uses is affected by urban infrastructure. For example, ground-level modifications should more strongly affect the airspace for species that fly close to the ground than for high-flying species [8]. Nonetheless, bats must navigate around anthropogenic elements, including densely arranged buildings of variable heights, windows, transmission lines, open roads, and vehicles, all of which increase collision risk and alter the structure of the airspace [5, 6]. Urban airspace is also affected by high noise levels, air pollution, and intense artificial light at night (ALAN) [4, 8]. Urban canyons (also called street canyons), created by clusters of tall buildings, can alter conditions in the atmospheric boundary layer by trapping noise and radiant heat, thereby altering wind speed and direction, temperature, and ultimately air quality [9]. Redirected echoes from the smooth surfaces and windows of buildings may also be disorienting to bats [10].

In this chapter, we explore how the structural and environmental modifications of the urban airspace might affect urban bats, based on existing literature that has tested the effects of these modifications in laboratory and field settings (non-urban and urban). We highlight key knowledge gaps associated with urban aeroecology of bats and, where sufficient evidence is available to inform potential solutions, propose actions to support bats that rely on urban aerial habitats.

2   Urban Modifications to Aerial Habitats

Urban bats must deal with a variety of anthropogenic modifications of the airspace, including dense roads with heavy traffic, intense anthropogenic noise (anthrophony) and artificial light at night (ALAN), heat islands, air pollution, street canyons that affect temperature and wind speed, and risk of collision with windows. Each of these modifications varies along a spectrum with increasing urban densification, and we consider them in detail below.

2.1 Roads and Vehicle Traffic

Given that roads and vehicle traffic often affect the movements of wildlife, including bats outside cities [11], how do they respond to roads and traffic in cities? Road density increases in urban environments, where roads and traffic range from two-lane, local roads with light traffic (e.g. residential areas) to multi-lane roads with heavy and fast-moving traffic (e.g. highways). Research outside cities suggests the contextual nature of bats' responses to roads and traffic – these responses seem to vary with road characteristics and surrounding habitat (e.g. number of lanes, traffic volume, vehicle noise, surrounding vegetation) and with variation among ecological traits among bat species. However, research in cities is sparse. Drawing on studies from non-urban areas, we consider how roads and traffic may affect urban bats and outline knowledge gaps to be addressed.

Bats may be less willing to cross wide roads compared to narrower, two-lane roads [11, 12], but roadside vegetation might mitigate road avoidance in cities by bridging the airspace and buffering disturbances from vehicle traffic below. Vegetation adjacent to roads can direct bats to cross where it is relatively safer, such as high above the road or through an underpass [12]. For example, acoustic monitoring of an urban population of northern myotis (*Myotis septentrionalis*) in Toronto, Canada, revealed that bats may use a forest corridor under a raised portion of a 14-lane highway to move between patches of preferred, interior habitat [13]. Bats in residential neighbourhoods in Toronto also cross tree-lined roads across the city (L. Moretto, pers. obs.). However, implementing roadside vegetation to guide crossings by bats should be well-planned because it can also raise the risk of collisions if it encourages low-altitude crossings [14]. Data from Brazil indicate that road crossings and vehicle strikes are most frequent where forest and other high-quality bat habitats intersect with a road, possibly because bats emerge from nearby roosts at lower altitudes [15].

Traffic noise and vehicle movements also generally disturb bats, interfering with their commuting and foraging [11, 12, 16–18]. However, most available studies have not been performed in urban settings, where bat responses may be distinct. In one laboratory study, greater mouse-eared bats (*Myotis myotis*) foraged less efficiently in response to playbacks of recorded traffic noise [16, 18]. Traffic noise also reduced general bat activity in another European study, but playback of ultrasonic sounds matching the frequencies of echolocation calls affected only one species, suggesting that traffic noise has additional effects beyond interfering with echolocation [17].

Demographic and functional traits of bats may partly predict their responses to roads and traffic and risk of vehicle strikes [11, 17, 19, 20]. Generally, collisions and road-related evasive manoeuvres increase during swarming, mating, and migration and for males and juveniles [12]. Species adapted to slow, low-altitude, manoeuvrable flight, such as those that glean or forage in clutter, may also be at a greater risk of collision because they often cross roads at lower altitudes and are less willing to make long-distance detours around roads [11, 17, 19, 21, 22]. For example, low-flying frugivorous and insectivorous phyllostomids were the most commonly detected roadkill in Brazil [15]. Conversely, some generalist species that are common in cities, such as pipistrelles (*Pipistrellus* spp.), do not exhibit road avoidance behaviours and thus may be at greater risk of collision with vehicles if they cross roads more often [23]. Unfortunately, with most knowledge of how traffic affects bats coming from non-urban contexts, it remains unclear whether urban bats, which may have habituated to constant traffic, might respond differently.

If some urban bats avoid road crossings, then how can we 'bridge the gap' to enhance connectivity across aerial habitat? Prioritising locations for 'enhanced connectivity' should start by mapping the movements of bats, e.g. using radiotelemetry to identify where bats actually cross roads (see [13, 22]). Once active crossing sites are discovered, vegetated over- or underpasses can be established to offer options for safer crossing [24]. For example, tall trees may help promote crossing over urban neighbourhoods or arterial roads at a safe height [12, 14], while crossing under elevated roads may be facilitated by continuous vegetation [13].

2.2   Urban Noise

Urban airspaces are acoustically complex and noisy, and the effects of this noise on bats are not clear. Some research suggests that bats might be able to tolerate urban noise. Most bats are sensitive to frequencies above 10 kHz, corresponding to calls used for social communication and ultrasonic calls used for echolocation [25]. Most anthropogenic sound in urban environments is loudest at frequencies below 2 kHz, and so is unlikely to interfere with activities that depend on ultrasonic sounds [26–28] or with high-frequency communication calls. Some bats tolerate and habituate to urban noise by adapting their activity to the general soundscape. For example, to deal with urban noise masking prey-generated sounds, some gleaning species may increase their use of visual cues for foraging [26, 27]. Bats in a noisy airspace may also adjust their calls; for example, Russo and Ancillotto [28] suggested that lowering the frequencies of social calls [e.g. by Kuhl's pipistrelles (*Pipistrellus kuhlii*)] in an urban area of Italy may reduce call attenuation, as lower-frequency sounds propagate further. In southern Australia, grey-headed flying foxes (*Pteropus poliocephalus*) appeared to simply tolerate urban noise – vocalisations were similar between urban and rural colonies [29].

Nevertheless, there may be thresholds of urban noise beyond which tolerance is limited. In the Australian study of *P. poliocephalus*, vocalisations at two highly urban roosts declined or ceased in response to loud noise produced by low-flying aircraft, despite these colonies' apparent tolerance for other urban noise and moving vehicles. In Chicago, United States, the activity of silver-haired (*Lasionycteris noctivagans*) and big brown bats (*Eptesicus fuscus*) declined in response to anthrophony [30]. Specific anthropogenic sounds associated with a greater risk of mortality, such as traffic noise, may alter bat behaviour [16, 18, 29]. Ultimately, we need to understand interspecific variation in bats' responses to anthropogenic noise in the urban airspace before we can develop effective mitigation strategies.

2.3   Urban Light Pollution

Artificial light at night (ALAN) is a recognised threat to global biodiversity [31] and affects roughly 23% of global land surface [32]. The effect of ALAN on bats varies with the type of light and species of bat [33–38]. Here, we summarise trends based on available evidence, which should be considered with the following two caveats. First, there is a pronounced latitudinal bias in research on ALAN [39], and most of the research on bat responses to ALAN has focused on temperate zone species. Second, although we include results from studies in rural areas, we acknowledge that bats accustomed to urban habitats may respond differently to ALAN.

The spectrum of ALAN is changing as lighting technology and consumer preferences evolve, and bat species appear to vary in their responses to different wavelengths and illuminance [38]. Older technologies, such as low-pressure sodium (LPS) and high-pressure sodium lights, release narrow-spectrum, yellow/orange light, whereas mercury vapour lights produce broader-spectrum, whiter light, including ultraviolet (UV) wavelengths. These are increasingly being replaced by broader-spectrum, light-emitting diodes (LEDs) and metal halide lights that also produce UV light [38, 40]. In European and British studies, most pipistrelles seem to exhibit higher activity at UV-emitting streetlights than at LED streetlights, whereas activity of *Myotis* spp. and parti-coloured bats (*Vespertilio murinus*) declines near ALAN regardless of wavelength [37, 41, 42]. Noctules (*Nyctalus* spp.), serotine bats (*Eptesicus serotinus*), and soprano pipistrelles (*P. pygmaeus*) exhibit inconsistent responses to different wavelengths of ALAN among studies [34, 37, 42, 43]. In contrast, tropical urban bats in Singapore exhibit no difference in activity between streets lit by HPS lights and streets retrofitted with LED lights [43]. Bat activity may also vary with the distribution of light in the airspace: activity of some species decreases more with vertical than with horizontal illuminance [36].

Artificial lights are associated with global insect declines [31] that pose a long-term threat to insectivorous bats, but lights can provide a foraging advantage for some bat species at the local scale by concentrating insects (e.g. [36, 44, 45]). Indeed, bat responses to different light spectra may partially reflect insect responses to the same lights. Lights that emit short wavelengths, particularly UV, typically attract more insects than lights with longer wavelengths (e.g. LPS streetlights; [31, 36]), so some lights provide better foraging opportunities than others [38]. This may be especially true for moths ([1, 46]; but see [47]).

Certain species are considered 'light-tolerant', but even they may modify their movements in response to ALAN [36, 38, 48]. *Pipistrellus pipistrellus* appears to tolerate lighting in rural areas where illumination is limited [48] but is less likely to leave tree cover to cross gaps in vegetation in cities, which are brightly lit and have large gaps [49]. Radio-tracked common noctules (*N. noctula*) in Berlin, Germany, tolerated ALAN when foraging in areas with abundant water or vegetation but avoided light when commuting between roosts and foraging areas [50]. In a park south of Paris, France, activity by *Pipistrellus* and *Nyctalus* spp. increased near streetlights, but the effect varied with illuminance [36], and both species preferred forest cover over open space as they approached lights, particularly white light [43]. Both fast- and slow-flying species of bats used lit streets to commute in Singapore, but feeding buzzes were rarely recorded, suggesting that the bats were foraging elsewhere [51].

Other species may restrict their activity to dark areas where possible. Forest bats in Sydney, Australia, exhibited higher activity within urban forests than near forest edges lit by streetlights [33]. In Europe, slow-flying, forest specialists [e.g. *Plecotus* sp., some *Myotis* sp., horseshoe bats (*Rhinolophus* spp.)], which are adapted to clutter, exhibit light avoidance behaviour, which may reflect a greater perceived risk of predation. Experiments in non-urban habitats revealed that *Myotis* and *Plecotus* sp. are less active when lights are on than when they are off [34, 35, 43], as also observed for *Myotis* bats in an urban habitat [37]. The effect of streetlights on *Myotis* activity extends up to 50 m from the lights, even at low illuminance (<1 lux), and persists after the lights are turned off [36]. Lesser horseshoe bats (*R. hipposideros*) also reduce activity and delay emergence in response to ALAN [35, 52, 53]. In Costa Rica, fast- and slow-flying urban bats both responded negatively to ALAN, suggesting light avoidance [54].

Solutions to mitigate the effects of ALAN on urban bats are similar to those proposed for other nocturnal species and require urban planners to explicitly account for biodiversity in land management and lighting decisions [48]. Existing lighting infrastructure can also be made more bat-friendly by decreasing the spectral intensity of LEDs and shifting to lower (yellower) spectra [55]. Of course, any modifications to street lighting must meet the perceived safety needs of humans. For example, new lights or retrofits that produce minimal vertical illuminance but enough horizontal illuminance for human use can benefit bats and still meet public safety requirements [21, 36].

2.4   Urban Heat Islands

Bats experience higher ambient temperatures in cities due to the urban heat island (UHI) effect, a phenomenon whereby cities are substantially warmer than surrounding, non-urban areas. Dark and impervious surfaces absorb radiation, and urban infrastructure traps heat, while remnant patches of vegetation continue to cool the air in specific locations, generating strong thermal gradients [56]. As a result, maximum air temperatures in some cities reach up to 15 °C higher than surrounding rural areas [57].

The interactive effects of global warming and UHIs on bats may be contextual. In the temperate zone, bats in UHIs might experience longer active seasons and warmer roosts than in non-urban areas. Warmer maternity roosts are associated with accelerated juvenile development in some species [58]. However, global warming may drive air and roost temperatures in some regions above the species' thermal tolerance [59, 60], resulting in heat stress and mortality. For example, extreme heat in the austral summer of 2019–2020 resulted in mass abandonment of pups at flying fox colonies in Australia and killed more than 72,000 individuals of various species [61]. Pteropodids also died of heat stress in and around the town of Pongola, South Africa [62]. These heat-related mortalities are not confined to cities, but UHIs certainly increase the frequency and severity of heatwaves [56].

Slowing climate change directly benefits bat populations everywhere, while urban bats will also benefit from urban planning to mitigate UHIs. Current methods include the use of permeable or reflective pavement, reflective building materials, and addition of green spaces and green infrastructure [57, 63].

2.5 Urban Canyons

From an urban bat's perspective, the airspace is characterised by frequent intrusion of buildings, especially in large urban cores, which often have densely packed skyscrapers [9]. In some ways, the urban canyons created by tall buildings mimic natural canyons; they provide roosting habitat analogous to that in rock crevices of escarpments. However, most natural canyons form over time by ephemeral or permanent rivers that buffer air temperature and provide access to drinking water and arthropod prey. In contrast, urban canyons rise from busy roads that present a collision risk to bats (although it is possible that bats can learn to avoid vehicle traffic more effectively in cities, given the more consistent volume of traffic in cities than non-urban areas).

High-velocity, turbulent airflow in urban canyons [56] may also pose a challenge to bats. As wind moves quickly above the top of clusters of tall buildings, some of the moving air is also pulled down into urban canyons, creating high spatial variation in wind speed, including downdraught and updraught [64] that could be difficult for a bat to navigate. We speculate that bats flying at high altitudes (e.g. during long-distance migrations) could be directed down into urban canyons by wind vortices they may not anticipate, but are not aware of any studies on the behavioural adaptations of flying wildlife to variable airflow in urban canyons. Even high-flying species that are well-adapted to navigating comparable wind speeds in natural canyons might struggle to navigate between buildings composed largely of smooth surfaces such as glass, concrete, or steel that they cannot grip when they require rest.

Urban canyons also contribute to UHI effects because wind vortexes trap heat and air pollution at street level [56], which can be mitigated by incorporating trees and other vegetation into street design. Hedges and other low vegetation can improve air quality at street level, and incorporating green walls can improve air quality at higher altitudes [9]. Tall street trees can also help cool urban canyons but, unlike hedges, can also concentrate air pollutants at street level as they trap some air below their canopy [65]. Implementing green roofs and walls on tall structures can improve aerial habitat quality [4] and provide shelter to bats foraging in or migrating through urban canyons. The effectiveness of these actions has not been evaluated, and post-construction monitoring should be conducted to identify and refine building designs that provide the greatest benefit to urban bats and other flying wildlife.

2.6 Windows

Windows are invisible barriers for flying wildlife (e.g. billions of birds in the United States die each year from window collisions; [66]), but the limited bat-related data suggests that windows are not a major threat to bats. Organisations that track wildlife window collisions in cities rarely report bat-window collisions. For example, at the time of this writing, the Toronto chapter of the Fatal Light Awareness Program has recorded >75,000 window strikes for birds since 2000 but <200 for bats (P. Plant, pers. comm., 2021). Similarly, since 2013, Lights Out Baltimore has reported only 85 bat strikes (L. Jacks, pers. comm., 2021), most during periods of migration.

Although observations of bats and window collisions to date have been passive or incidental and windows may not pose a substantial threat to bats, a series of intriguing experiments is revealing how flying bats may perceive windows. At an obtuse angle of approach, a window's smooth surface redirects most of a bat's echolocation calls away from the bat, which could confuse the bat [10]. Thus, angle of approach may influence the likelihood of collision. However, bats use visual cues alongside acoustic cues to improve perception [67, 68], which may help bats to perceive windows in their flight path and avoid collisions. Unless further data on bat-window collisions arise to suggest otherwise, it seems unlikely that window collisions pose a real threat to urban bats.

3   Bat-Friendly Stewardship of Urban, Aerial Habitats

Although we have presented potential actions to support urban bats throughout the chapter, these require robust evaluation before they can be recommended to policy-makers, urban planners, and conservation practitioners. We had to draw on many non-urban studies, but city-dwelling bats may be habituated to urban aerial habitat modification and respond differently. Accurately generalising the needs of bats in urban, aerial habitats is also limited by high interspecific variation in bats' responses to the risk of vehicle strikes, habitat modification by roads, urban noise and light pollution, and tolerance of extreme heat. Nevertheless, the potential solutions we have summarised here overlap with urban habitat management that supports other urban wildlife and human wellbeing.

Evidence-based stewardship projects to support aerial habitats for other urban species can be leveraged to test effects on urban bats. Urban planners, developers, and researchers should collaborate to enable robust experimental designs such as before-after-control-impact studies to quantify bat responses to habitat management. Reducing ALAN where possible and incorporating vegetation into urban infrastructure provide habitat for other wildlife and an opportunity to test how bats respond to these changes in aerial commuting pathways and foraging habitat. More specifically, reducing high-intensity, vertical lighting where bats cross roads may work best when combined with tree planting to reduce perceived gap width. Finally, any actions to slow climate change are likely to benefit urban bats by reducing UHI effects and canyon effects.

Acknowledgements We are grateful to Brendan Samuels (Western University), Paloma Plant (Fatal Light Awareness Program Canada), and Lindsay Jacks (Lights Out Baltimore) for sharing information about bat collisions with windows and Allison Fairbrass for helpful discussions about anthropogenic noise. Two anonymous reviewers and the editor of this chapter provided helpful comments on an earlier version of this chapter.

Literature Cited

1. Diehl RH (2013) The airspace is habitat. Trends Ecol Evol 28:377–379
2. Kunz TH et al (2008) Aeroecology: probing and modeling the aerosphere. Integr Comp Biol 48:1–11
3. Chilson PB et al (2018) Aeroecology. Springer, Cham
4. Zuluaga S et al (2021) Global aerial habitat conservation post-COVID-19 anthropause. Trends Ecol Evol 36:273–277
5. Lambertucci SA, Speziale KL (2020) Need for global conservation assessments and frameworks to include airspace habitat. *Conserv Biol.* https://doi.org/10.1111/cobi.13641
6. Davy CM et al (2017) Aeroconservation for the fragmented skies. Conserv Lett 10:773–780
7. Voigt CC et al (2018) Conservation strategies for bats flying at high altitudes. Bioscience 68:427–435
8. Diehl RH et al (2018) Extending the habitat concept to the airspace. In: Chilson PB et al (eds) Aeroecology. Springer, Cham, pp 47–69
9. Abhijith KV et al (2017) Air pollution abatement performances of green infrastructure in open road and built-up street canyon environments – a review. Atmos Environ 162:71–86
10. Greif S et al (2017) Acoustic mirrors as sensory traps for bats. Science 80:1045–1047
11. Kitzes J, Merenlender A (2014) Large roads reduce bat activity across multiple species. PLoS One 9:e96341
12. Fensome AG, Mathews F (2016) Roads and bats: a meta-analysis and review of the evidence on vehicle collisions and barrier effects. Mamm Rev 46:311–323
13. Thorne TJ et al (2021) Occurrence of a forest-dwelling bat, northern myotis (*Myotis septentrionalis*), within Canada's largest conurbation. J Urban Ecol 7:1–9
14. Ramalho DF, Aguiar LMS (2021) Bats on the road – a review of the impacts of roads and highways on bats. Acta Chiropterologica 22:417–433
15. Novaes RLM et al (2018) On a collision course: the vulnerability of bats to roadkills in Brazil. Mastozoología Neotrop 25:115–128
16. Schaub A et al (2008) Foraging bats avoid noise. J Exp Biol 211:3174–3180
17. Finch D et al (2020) Traffic noise playback reduces the activity and feeding behaviour of free-living bats. Environ Pollut 263:114405
18. Siemers BM, Schaub A (2011) Hunting at the highway: traffic noise reduces foraging efficiency in acoustic predators. Proc R Soc B Biol Sci 278:1646–1652

19. Claireau F et al (2019) Major roads have important negative effects on insectivorous bat activity. Biol Conserv 235:53–62
20. Jung K, Threlfall CG (2015) Urbanisation and its effects on bats—a global meta-analysis. In: Voigt CC, Kingston T (eds) Bats in the anthropocene: conservation of bats in a changing world. Springer, Cham, pp 1–606
21. Altringham J, Kerth G (2016) Bats and roads. In: Bats in the anthropocene: conservation of bats in a changing world. Springer, Cham, pp 35–62
22. Claireau F et al (2019) Bat overpasses: an insufficient solution to restore habitat connectivity across roads. J Appl Ecol 56:573–584
23. Medinas D et al (2013) Assessing road effects on bats: the role of landscape, road features, and bat activity on road-kills. Ecol Res 28:227–237
24. Kerth G, Melber M (2008) Species-specific barrier effects of a motorway on the habitat use of two threatened forest-living bat species. Biol Conserv 142:270–279
25. Götze S et al (2020) High frequency social calls indicate food source defense in foraging common pipistrelle bats. Sci Rep 10:1–9
26. Moretto L, Francis CM (2017) What factors limit bat abundance and diversity in temperate, North American urban environments? J Urb Ecol 3:1–9. https://doi.org/10.1093/jue/jux016
27. Gomes DGE et al (2016) Bats perceptually weight prey cues across sensory systems when hunting in noise. Science 353:1277–1280
28. Russo D, Ancillotto L (2015) Sensitivity of bats to urbanization: a review. Mamm Biol 80:205–212
29. Pearson T, Clarke JA (2019) Urban noise and grey-headed flying-fox vocalisations: evidence of the silentium effect. Urban Ecosyst 22:271–280
30. Lehrer EW et al (2021) Urban bat occupancy is highly influenced by noise and the location of water: considerations for nature-based urban planning. Landsc Urban Plan 210:104063
31. Owens AC et al (2020) Light pollution is a driver of insect declines. Biol Conserv 241:108259
32. Falchi F et al (2016) The new world atlas of artificial night sky brightness. Sci Adv 2:e1600377
33. Haddock JK et al (2019) Light pollution at the urban forest edge negatively impacts insectivorous bats. Biol Conserv 236:17–28
34. Spoelstra K et al (2017) Response of bats to light with different spectra: light-shy and agile bat presence is affected by white and green, but not red light. Proc R Soc B Biol Sci 284:20170075
35. Zeale MRK et al (2018) Experimentally manipulating light spectra reveals the importance of dark corridors for commuting bats. Glob Chang Biol 24:5909–5918
36. Azam C et al (2018) Landscape and urban planning evidence for distance and illuminance thresholds in the effects of artificial lighting on bat activity. Landsc Urban Plan 175:123–135
37. Straka TM et al (2019) Tree cover mediates the effect of artificial light on urban bats. Front Ecol Evol 7:91
38. Rowse EG et al (2016) Dark matters: the effects of artificial lighting on bats. In: Bats in the anthropocene: conservation of bats in a changing world. Springer, Cham, pp 187–214
39. Secondi J et al (2020) Assessing the effects of artificial light at night on biodiversity across latitude – current knowledge gaps. Glob Ecol Biogeogr 29:404–419
40. Elvidge CD et al (2010) Spectral identification of lighting type and character. Sensors 10:3961–3988
41. Lewanzik D, Voigt CC (2017) Transition from conventional to light-emitting diode street lighting changes activity of urban bats. J Appl Ecol 54:264–271
42. Stone EL et al (2015) The impacts of new street light technologies: experimentally testing the effects on bats of changing from low- pressure sodium to white metal halide. Philos Trans B 370:20140127
43. Barré K et al (2021) Bats seek refuge in cluttered environment when exposed to white and red lights at night. Mov Ecol 9:1–11
44. Hickey MBC et al (1996) Resource partitioning by two species of vespertilionid bats (*Lasiurus cinereus* and *Lasiurus borealis*) feeding around street lights. J Mammal 77:325–334

45. Acharya L, Fenton M (1999) Bat attacks and moth defensive behaviour around street lights. Can J Zool 77:27–33
46. Van Langevelde F et al (2011) Effect of spectral composition of artificial light on the attraction of moths. Biol Conserv 144:2274–2281
47. Law B et al (2019) Responses of insectivorous bats and nocturnal insects to local changes in street light technology. Austral Ecol 44:1052–1064
48. Pauwels J et al (2019) Landscape and urban planning accounting for artificial light impact on bat activity for a biodiversity- friendly urban planning. Landsc Urban Plan 183:12–25
49. Hale JD et al (2015) The ecological impact of city lighting scenarios: exploring gap crossing thresholds for urban bats. Glob Chang Biol 21:2467–2478
50. Voigt CC et al (2020) Movement responses of common noctule bats to the illuminated urban landscape. Landsc Ecol 4:189–201
51. Lee KEM et al (2021) Ecological impacts of the LED-streetlight retrofit on insectivorous bats in Singapore. PLoS One 16:1–14
52. Stone EL et al (2009) Street lighting disturbs commuting bats. Curr Biol 19:1123–1127
53. Stone EL et al (2012) Conserving energy at a cost to biodiversity? Impacts of LED lighting on bats. Glob Chang Biol 18:2458–2465
54. Frank TM et al (2019) Impact of artificial lights on foraging of insectivorous bats in a Costa Rican cloud forest. J Trop Ecol 35:8–17
55. Kerbiriou C et al (2020) Switching LPS to LED streetlight may dramatically reduce activity and foraging of bats. Diversity 12:1–14
56. Oke TR et al (2017) Urban climates. Cambridge University Press, Cambridge, p 456
57. Aflaki A et al (2017) Urban heat island mitigation strategies: a state-of-the-art review on Kuala Lumpur, Singapore and Hong Kong. Cities 62:131–145
58. Zahn A et al (2007) Critical times of the year for *Myotis myotis*, a temperate zone bat: roles of climate and food resources. Acta Chiropterologica 9:115–125
59. Crawford RD, O'Keefe JM (2021) Avoiding a conservation pitfall: considering the risks of unsuitably hot bat boxes. Conserv Sci Pract. https://doi.org/10.1111/csp2.412
60. Martin Bideguren G et al (2019) Bat boxes and climate change: testing the risk of over-heating in the Mediterranean region. Biodivers Conserv 28:21–35
61. Mo M et al (2021) Estimating flying-fox mortality associated with abandonments of pups and extreme heat events during the austral summer of 2019–20. Pacific Conserv Biol. https://doi.org/10.1071/PC21003
62. McKechnie AE et al (2021) Mortality among birds and bats during an extreme heat event in eastern South Africa. Austral Ecol 46:687–691
63. Wang C et al (2021) Cool pavements for urban heat island mitigation: a synthetic review. Renew Sust Energ Rev 146:111171
64. Walker SL (2011) Building mounted wind turbines and their suitability for the urban scale—a review of methods of estimating urban wind resource. Energ Buildings 43:1852–1862
65. Coutts AM et al (2016) Temperature and human thermal comfort effects of street trees across three contrasting street canyon environments. Theor Appl Climatol 124:55–68
66. Loss SR et al (2014) Bird–building collisions in the United States: estimates of annual mortality and species vulnerability. Condor 116:8–23
67. Orbach D, Fenton MB (2010) Vision impairs the abilities of bats to avoid colliding with stationary obstacles. PLoS One 5:e13912
68. Geipel I et al (2019) Noise as an informational cue for decision-making: the sound of rain delays bat emergence. J Exp Biol. https://doi.org/10.1242/jeb.192005

Chapter 8
City Trees, Parks, and Ponds: Green and Blue Spaces as Life Supports to Urban Bats

Lauren Moretto, Leonardo Ancillotto, Han Li, Caragh G. Threlfall, Kirsten Jung, and Rafael Avila-Flores

Abstract Patches of vegetated habitat within urban areas ("green spaces") and water bodies ("blue spaces") are crucial to support urban wildlife, including bats. In this chapter, we review the literature to explore how bats use green and blue spaces, including natural, semi-natural, and manicured vegetated areas, and various water bodies. We first examine the value of urban green spaces to bats for roosting, foraging, commuting, and refuge from disturbances. We then examine the importance of blue spaces as sources of drinking water and prey. We also consider how spatial arrangements of green and blue spaces across the urban landscape influence use by bats. Finally, we review approaches of studying green and blue spaces to guide future research and suggest guidelines for better design and management of these valuable habitats to support urban bat abundance and diversity.

L. Moretto (✉)
Vaughan, ON, Canada

L. Ancillotto
Wildlife Research Unit, Dipartimento di Agraria, Università degli Studi di Napoli Federico II, Naples, Italy

H. Li
Department of Biology, University of North Carolina at Greensboro, Greensboro, NC, USA

Department of Biology, University of Nebraska Omaha, Omaha, NE, USA

C. G. Threlfall
School of Life and Environmental Science, The University of Sydney, Sydney, NSW, Australia

K. Jung
Institute of Evolutionary Ecology and Conservation Genomics, University Ulm, Ulm, Germany

R. Avila-Flores
División Académica de Ciencias Biológicas, Universidad Juárez Autónoma de Tabasco, Villahermosa, Tabasco, Mexico

Keywords Green space · Blue space · Foraging · Drinking · Habitat · Urban landscape

1 Introduction

Urbanisation fundamentally alters natural environments around the world. Urban areas become a mosaic of impervious surfaces with patches of semi-natural, vegetated habitat when large, interconnected patches of habitat are transformed with artificial structures. These remaining patches, hereafter "green spaces" [1], include urban nature reserves and parks, street trees, private lots, and community gardens (Figs. 8.1 and 8.2). However, they are mostly small, fragmented, heavily manicured, and dominated by non-native vegetation [1–3] (Fig. 8.2).

Although these green spaces are often heavily transformed, they remain valuable to wildlife, including bats [4–9]. In this chapter, we explore how various green spaces may be used by bats in urban environments. In particular, we examine how green spaces provide critical resources and spaces for roosting, foraging, commuting, and refuge from disturbance. We also investigate the importance of waterbodies

Fig. 8.1 Green spaces provide important resources to support populations of bats in urban areas, including spaces for roosting, foraging, commuting, and hiding from predators. Large green spaces, like this urban park in Ulm, Germany, may support the diversity of bats in cities by providing habitat for urban-sensitive species

Fig. 8.2 Many green spaces in urban areas are small in size, but nonetheless, they provide valuable habitat for urban bats. A city with many small fragments of green space may help to distribute resources for bats across the landscape

within these green spaces, or "blue spaces", to bats. We then consider the influence of spatial, landscape-scale arrangements of green and blue spaces on the use by bats. Lastly, we review approaches to studying urban green and blue spaces to guide future research, and we provide recommendations to improve management and design of urban green and blue spaces to support urban bat populations.

We recognize that bats will perceive green and blue spaces differently from humans. Understanding this perspective from a sensory and experiential point of view is a challenging task (especially since bats rely primarily on auditory perception to experience the world, unlike humans [10]). For example, humans may consider street trees, private gardens, and yards as being different types of green spaces. Bats may perceive these spaces as general habitat patches with different purposes (e.g. foraging and roosting) or levels of quality/value (e.g. a street tree may have no value to a bat due to surrounding disturbance, while trees in parks may provide safe roosting space). The same "type" of green or blue space (e.g. all street trees) may also be perceived differently across an urban environment. For example, street trees that are within or close to other roosting, foraging, or drinking locations may be preferred to street trees which are further from these sites [11]. The quality, availability, and functions of resources within green and blue spaces may differ temporally (e.g. seasonally), which may alter how bats will use these spaces over time. For instance, trees that are used for summer roosts may not provide adequate roosting space in the winter months, and tree hollows that are used as maternity roosts may differ from those used outside of this period [6, 12]. Ultimately, we recognize that our understanding of bat needs and preferences in urban areas can easily be biased by human perception. Using existing and emerging knowledge about bat ecology and behaviour will help to remove human biases. This will benefit our ability to understand how bats use the urban environment and better inform management decisions for bats.

2   Use of Green Spaces by Urban Bats

Urban green spaces vary in their composition (e.g. vegetation types and structure), management, and use by humans, but research suggests that these green spaces provide critical resources for bats in urban environments. Below, we describe how urban bats may use these green spaces.

2.1   Roosting Habitat

Green spaces provide vegetation for roosting and improve the quality of artificial roosts nearby (e.g. bat boxes). For tree-roosting bats, trees with nooks, crevices, and large crowns (e.g. old trees or dead trees/snags) provide several spaces for roosting [6, 11–13]. Unfortunately, these trees are often removed in city centres for aesthetic or safety reasons (i.e. falling branches/trunks may pose a hazard to people if a tree is close to streets, public paths and trails, buildings, etc.), but they may be retained in parks and nature reserves. These larger green spaces have several diverse roosting structures and can support populations of different bat species [7, 8, 14–16]; both common and rare bat species have been recorded roosting in urban green spaces [6, 11–13]. Bats that roost in artificial structures, like bat boxes and houses, also prefer roosting near green spaces. For example, Kubista et al. found that roost site selection in human houses was positively influenced by the amount of green space nearby [16]. Green spaces may improve the quality of these artificial roosts by providing easy access to other resources, such as food and water (see Sect. 4 below).

2.2   Foraging Habitat

Green spaces provide food and foraging spaces for urban bats. Frugivorous and nectivorous bat species feed on the nectar and fruits from trees and other plants in urban green spaces [17–20]. In Tuxtla Gutiérrez, southern Mexico, the relative abundance (number of individuals captured in ground-level mist nets) of Jamaican fruit bats (*Artibeus jamaicensis*) in urban parks was positively influenced by the abundance and quality of fruits in these parks; in particular, quantity of nitrogen available in fruits was the best predictor of bat abundance [21]. Vegetation in green spaces also supports populations of prey for insectivorous bats [8, 22, 23]. In urban

Sydney, Australia, green patches and backyards in areas of the city with nutrient-rich soils and moderate vegetation cover supported greater insect biomass and significantly greater bat feeding activity than other parts of the city with nutrient-poor soils and less vegetation [23]. Similarly, in urban Melbourne, Australia, a positive relationship was observed between vegetation cover within green spaces, abundance of nocturnal invertebrates, and activity and species richness of insectivorous bats [24]. Researchers also speculate that green roofs may provide important foraging spaces for insectivorous bats, since they support more insect populations and bat activity than conventional roofs [25–27].

2.3   Corridors for Commuting and Refuge from Disturbances

Green spaces provide commuting routes for bats and connect roosting and foraging areas in urban environments. For example, linked tree canopies may help bats commute across urban areas. Studies have recorded bats flying along tree canopies in urban areas, possibly because they physically bridge habitat, orient commuting bats, and mitigate disturbances by reducing exposure to artificial lighting and buffering anthropogenic noise [11, 13, 28–37]. However, not all bat species will use these commuting routes equally; bats may be selective of routes, depending on their tolerance to anthropogenic disturbance. For example, bat species that opportunistically forage around street lights [e.g. great fruit-eating bat (*Artibeus lituratus*)] may prefer using tree-lined streets as commuting routes while disturbance-sensitive species may not [11, 28].

Large urban green spaces may provide refuge from anthropogenic disturbances for bats. A study of urban forest patches by Haddock et al. found that the dark interior of urban forest patches provided habitat for light-sensitive species [29]. In Greensboro, North Carolina, USA, more bat activity was found in peripheral city parks with less human activity than central city parks with more human activity [35]. In Melbourne, Australia, urban golf courses in areas of moderate anthropogenic activity were extremely important for supporting bat species richness. These golf courses contained large, old trees, understory vegetation that supported insect populations, water bodies, and large expanses of dark spaces for foraging [24].

3   Blue Spaces in and around Urban Green Spaces

Urban green spaces often contain blue spaces, such as ponds, streams, fountains, and artificial pools, which may provide critical resources for urban bats. Despite being common in urban environments worldwide, their importance to urban bats has historically been under-examined, but research about blue spaces is increasing [37–40]. Studies suggest that urban green spaces that contain or are found near blue spaces support more bat activity when compared to green spaces without or far from blue spaces [41–44]. Furthermore, characteristics of blue spaces, including water availability, water quality, and the quality of green space surrounding the water source, may also influence bats [40–42]. Given this recent research into blue spaces, we examine the possible value of blue spaces to urban bats.

The primary value of blue spaces to bats is their provision of drinking water, especially during periods of drought. Bats require regular access to drinking water due to their morphology (i.e. small size and high surface-to-volume ratio of their wing membranes), which makes them susceptible to dehydration. Drinking involves strategic manoeuvring to lap water from the surface, so easily accessible and unclut-tered blue spaces (i.e. easier to perceive returning echolocation calls) are often preferred drinking sites (Fig. 8.3). Many urban blue spaces, like ornamental/recreational ponds and swimming pools, are often free of obstacles and may provide easy access to drinking water for bats. Bats also prefer drinking from water with fewer or less potent chemical treatments (e.g. swimming pools with mineral treatments are

Fig. 8.3 Blue spaces in urban environments provide sites for foraging and critical sources of drinking water for bats in urban environments. This brown long-eared bat (*Plecotus auratus*) from Italy manoeuvres close to the water surface to lap a drink with its tongue. Uncluttered blue spaces can improve access to drinking water (Photo credit to Sherri and Brock Fenton)

preferred to those with chlorine and salt [45]), though research has yet to address the effects of these chemicals on urban bats.

Blue spaces are also often structurally complex and seasonally variable, providing a variety of unique resources for urban bats. Several blue spaces in urban environments feature concrete embankments and culverts to direct water flow or prevent flooding, which may provide artificial roosting habitat for bats [46, 47]. Blue spaces in recreational areas or private gardens are often surrounded by vegetation or contain aesthetic features, such as fountains and waterfalls, which provide spaces and resources for foraging.

Blue spaces are important foraging spaces to some species of bats. Insectivorous bats like big brown bats (*Eptesicus fuscus*) and Mexican free-tailed bats (*Tadarida brasiliensis*) often forage for insect prey above water [8], and blue spaces may support nectar-yielding flowers for nectivorous bat species. In Melbourne, Australia, Straka et al. found that water bodies (including ponds) in green spaces supported abundant populations of nocturnal flying insects, including flies (Diptera) and caddisflies (Trichoptera). As a result, these spaces supported greater bat activity than similarly sized green spaces without blue spaces [42]. Blue spaces may also concentrate prey populations for bat species that glean prey from vegetation by hosting juvenile stages of insects and amphibians on aquatic vegetation. For example, urban populations of túngara frogs (*Engystomops pustulosus*) gather at ponds within green spaces for reproduction, concentrating these population for the frog-eating, fringe-lipped bat (*Trachops cirrhosus*) in urban areas [48, 49]. However, urban blue spaces that are highly managed or treated with insecticides may reduce the availability and quality of prey items.

Bats may prefer particular characteristics of blue spaces. Generally, bat activity is greater around larger blue spaces [41, 42] and around blue spaces with more riparian vegetation [39, 40, 42]. Irrespective of the size and presence of natural features around blue spaces, artificial illumination generally negatively impacts bat activity in and around blue spaces [40, 41, 50, 51]. Water quality and pollution may also influence the use of urban blue spaces by bats (e.g. [45]), but more research about this is required to better understand its effects on bats. Straka et al. found that bat activity was notably greater at ponds with lower levels of heavy metals in Melbourne [41], but Naidoo et al. [52] and Kalcounis-Rueppel et al. [53] found that eutrophic rivers host greater bat richness and activity levels when compared to nearby unpolluted urban rivers.

4   Spatial Considerations for Green and Blue Spaces in the Urban Landscape

The spatial organization of patches of green and blue space across an urban landscape influences a bat's use of these spaces [7, 32, 54]. Specifically, bats can be influenced by the total amount of green and blue space within an urban area (i.e. composition of patches in an urban area) and the spatial arrangement of these green and blue spaces (i.e. configuration of patches in an urban area; see [55, 56]). We explore the effects of both urban green and blue space composition and configuration on bats in this section.

Green and blue space composition in urban environments influences urban bat activity, abundance, and diversity. Simply, more green or blue space will provide more resources and potentially easier access to resources (e.g. prey refuge areas). More resources may attract and/or support larger populations of bats and a greater diversity of species. Research suggests that larger green spaces in urban areas, including large urban parks or golf courses, support greater abundance and diversity of bats than small green spaces [4, 8, 12, 15, 57, 58]. Although most of these studies examined the effects of large, unfragmented patches of habitat on bats, a large amount of fragmented habitat across an urban landscape may still importantly support bat populations [4, 8, 9, 11, 15].

Recently, researchers are examining the effects of fragmentation alone on urban bats (e.g. the value of a few large green space patches is compared to several smaller green space patches with the same total area). For some bat species, if the quality of many small, fragmented green space patches is high, these patches may contain enough high-quality resources across an urban area to support populations of these species [4, 8, 9, 11, 15, 59]. For example, highly mobile insectivorous species, like molossids, may find that many small, fragmented green spaces distribute pockets of prey across an urban landscape ([8]; Fig. 8.2). This may improve access to prey and foraging success. Conversely, other species may require large, unfragmented green spaces ([59, 60]; Fig. 8.1). For example, in Melbourne, Australia, Caryl et al. found that many bat species were no longer present in landscapes with medium densities of houses, irrespective of the total amount of tree cover across the landscape [61]. Therefore, they suggest that fewer large patches of green space may provide better conservation outcomes for bats in Melbourne than many smaller patches. If this is not achievable, they recommend that at least 20% of the landscape contain trees, as below this threshold in tree cover many species no longer occurred.

The configuration, or spatial arrangements, of green and blue spaces across an urban area also influences species diversity. The presence of a variety of complementary habitat types in an urban environment could suit the needs of a variety of species, thereby supporting greater species diversity. For example, Ancillotto et al. investigated the use of a pond archipelago by an assemblage of four generalist pipistrelle species, namely, common pipistrelle (*Pipistrellus pipistrellus*), soprano pipistrelle (*P. pygmaeus*), Kuhl's pipistrelle (*P. kuhlii*), and Savi's pipistrelle (*Hypsugo savii*), in Rome, Italy [40]. They found that all bats were present at ponds with woodland less than 1 km away and at ponds adjacent to woodland margins and hedgerows. The close proximity of complementary habitat types in this study may have reduced commuting times between roosts to foraging locations, risk of predation, and exposure to unfavourable weather conditions (e.g. wind, rain). Straka et al. also found that the amount of vegetation cover in the surrounding 0.5 km [measured using remote sensing as mean normalized difference vegetation index (NDVI) in the buffer area] strongly influenced the use of blue spaces by bats [42]. Inversely, there is also evidence suggesting that the absence of complementary habitat types in close proximity negatively affects bats; Lintott et al. found that greater amounts of built-up area surrounding an urban river negatively affected bat activity up to 3 km away [39].

5   The Study of Urban Green and Blue Spaces

Continuing to study urban green and blue spaces and their characteristics is important to guide future conservation efforts for bats. A common approach to studying the influence of these spaces involves surveying bat activity and species richness within different land cover types (e.g. commercial areas, residential areas, large parks, etc.) across urban environments (e.g. [4, 5, 7, 14, 15, 54]). Results from these studies may identify green and blue spaces that are most used by species of interest (e.g. [13, 58, 60]), which can inform management strategies in an urban area.

When green and blue spaces with high bat activity or species richness are identified, studies often examine characteristics of these spaces in detail to better understand bat habitat and resource needs (e.g. proximity of roost sites to feeding habitats, food availability, or human disturbance). For example, species not commonly found in urban areas can be found in large urban parks and nature reserves [12, 57], so characteristics of these green spaces are closely examined (e.g. light levels, water quality, tree cover, etc.), and the movements and behaviours of bats across these spaces are tracked [13, 14, 18, 19, 30, 35]. Alternatively, comparing the characteristics of green/blue spaces to those of non-green/blue space sites may also help to highlight characteristics of importance to urban bats (e.g. building design, water courses adjacent to wooded areas, parks without artificial lighting, etc.; see [8, 9]).

Observations of bats using other green or blue spaces not previously considered, including urban street treelines, private lots, green infrastructure (e.g. green roofs), urban water ways, and swimming pools, are inspiring research into characteristics of these spaces and their resources [11, 15, 25–27, 39, 41, 62]. For example, studies from Australia found that many species use residential gardens and lots, but less so than other larger or more vegetated spaces [24, 54, 63]. Brownfields (i.e. previously developed and abandoned sites) may also be used as foraging sites for insectivorous species, since they may support insect populations, but merit further examination as they retain pollutants which may harm the health of bats [52, 53]. Although some of these spaces may be challenging to access for research, such as private lots and residential properties, they should not be disregarded as they too provide important resources for urban bats [15, 64, 65].

6 Recommendations for Green and Blue Space Design and Management to Support Urban Bats

Here, we provide general recommendations for urban green and blue space design and management to support urban bat populations based on existing knowledge and expert opinion. However, these recommendations should not replace consideration of local context and site-/species-specific studies, which will always provide the best guidance for management action. The use of these spaces by bats is still not completely understood and will vary between urban environments around the world:

1. Green and blue spaces can provide resources for roosting, foraging, and drinking, corridors for commuting, and refuge from disturbance. Large green and blue spaces with a diversity of features (e.g. vegetation of various heights, dense forest, open spaces, dark spaces, water bodies surrounded by vegetation) should be prioritized for protection, as they may be able to support greater bat activity and species richness in urban areas. Specifically, the following features may help to promote urban bat diversity in green and blue spaces:

   • Patches of "interior" habitat (i.e. dark, buffered from anthropogenic disturbances such as light and noise, and highly vegetated) may support species that are less tolerant to disturbance (e.g. edge-sensitive or forest-interior species).
   • Large, older trees should be protected within green spaces when possible (balanced with removal of hazardous branches or trees for human safety) since they may provide support to populations of tree-roosting species in urban environments. Large trees provide several nooks and crevices for roosting and often harbour fruits or prey.
   • Blue spaces should contain both uncluttered areas for easier drinking and be surrounded by vegetation to concentrate prey for bats. Reducing artificial illumination in these spaces may also promote bat activity.

2. Green and blue space use by bats and associated management efforts should be considered across an urban landscape (not only within a single green or blue space patch):

   • The greatest total amount of green and blue space should be protected in an urban area. If large, unfragmented green and blue spaces cannot be protected, several smaller but high-quality patches should be protected.
   • Spaces may provide different functions and resources to bats (which may change seasonally), so a variety of spaces should be close together to improve access to these resources (e.g. within 1 km; see [15, 40, 58, 60]). For example, green spaces should contain or be close to blue spaces. This will improve access to drinking water and may concentrate prey items for urban bats.

7 Future Research Directions

There is still much to learn about how urban bats use green and blue spaces. Here, we highlight key knowledge gaps to guide future directions for research:

1. The use of green and blue spaces in private, residential lots should be studied when possible to better understand how these spaces support bat populations across an urban landscape (see [15, 64, 65]). For example, while residential lots may provide important spaces and resources for foraging or commuting, their value may depend on their proximity to larger patches of roosting habitat or on the degree of disturbances or threats present at these sites (e.g. backyard lights, cats and other predators, use of pesticides by residents).
2. The use of brownfields by bats should be explored further. Brownfields may be used as foraging sites for insectivorous species since they may support insect populations. However, the concentration of pollutants in brownfields may harm the health of bats. For instance, bats may consume insects with high concentrations of pollutants or drink from polluted water at these sites.
3. Blue spaces can be highly important to urban bats, but there are still many unanswered questions about their influences, including:

    - How valuable are blue spaces such as fountains and private swimming pools to bats? These are common in urban areas, but their use by bats is still poorly understood [45, 66]. Their function as sources of water for bats should be explored further given their permanence, especially to urban bat populations during dry seasons and/or in semiarid environments [66, 67].
    - What is the effect of chemically treated or polluted blue spaces (e.g. chlorinated swimming pools, polluted storm water ponds) on bat health in the short and long term?

Acknowledgments We are grateful to the editor of this chapter and two anonymous reviewers for the helpful comments and suggestions for revisions.

Literature Cited

1. Aronson MFJ, Lepczyk CA, Evans KL, Goddard MA, Lerman SB, MacIvor JS et al (2017) Biodiversity in the city: key challenges for urban green space management. Front Ecol Environ [Internet] 15(4):189–196. https://doi.org/10.1002/fee.1480
2. Niemelä J (1999) Ecology and urban planning. Biodivers Conserv [Internet] 8(1):119–131. https://doi.org/10.1023/A:1008817325994
3. Donihue CM, Lambert MR (2015) Adaptive evolution in urban ecosystems. Ambio [Internet] 44(3):194–203. https://doi.org/10.1007/s13280-014-0547-2
4. Threlfall CG, Law B, Banks PB (2012) Sensitivity of insectivorous bats to urbanization: implications for suburban conservation planning. Biol Conserv [Internet]. 146(1):41–52. Available from: https://www.sciencedirect.com/science/article/pii/S0006320711004459

5. Suarez-Rubio M, Ille C, Bruckner A (2018) Insectivorous bats respond to vegetation complexity in urban green spaces. Ecol Evol [Internet] 8(6):3240–3253. https://doi.org/10.1002/ece3.3897

6. Threlfall CG, Law B, Banks PB (2013) Roost selection in suburban bushland by the urban sensitive bat Nyctophilus gouldi. J Mammal [Internet] 94(2):307–319. https://doi.org/10.1644/11-MAMM-A-393.1

7. Li H, Wilkins KT (2014) Patch or mosaic: bat activity responds to fine-scale urban heterogeneity in a medium-sized city in the United States. Urban Ecosyst [Internet]. 17(4):1013–1031. https://doi.org/10.1007/s11252-014-0369-9

8. Avila-Flores R, Fenton MB (2005) Use of spatial features by foraging insectivorous bats in a large urban landscape. J Mammal [Internet]. 86(6):1193–1204. https://doi.org/10.1644/04-MAMM-A-085R1.1

9. Silva de Araújo MLV, Bernard E (2016) Green remnants are hotspots for bat activity in a large Brazilian urban area. Urban Ecosyst [Internet] 19(1):287–296. https://doi.org/10.1007/s11252-015-0487-z

10. Nagel T (1974) What is it like to be a bat? Philos Rev 83:435–450

11. Oprea M, Mendes P, Vieira TB, Ditchfield AD (2009) Do wooded streets provide connectivity for bats in an urban landscape? Biodivers Conserv [Internet]. 18(9):2361–2371. https://doi.org/10.1007/s10531-009-9593-7

12. Moretto L, Francis CM (2017) What factors limit bat abundance and diversity in temperate, North American urban environments? J Urban Ecol [Internet]. 3(1):jux016. https://doi.org/10.1093/jue/jux016

13. Dietz M, Bögelsack K, Krannich A, Simon O (2020) Woodland fragments in urban landscapes are important bat areas: an example of the endangered Bechstein's bat Myotis bechsteinii. Urban Ecosyst [Internet]. 23(6):1359–1370. https://doi.org/10.1007/s11252-020-01008-z

14. Gallo T, Lehrer EW, Fidino M, Kilgour RJ, Wolff PJ, Magle SB (2018) Need for multiscale planning for conservation of urban bats. Conserv Biol [Internet] 32(3):638–647. https://doi.org/10.1111/cobi.13047

15. Moretto L, Fahrig L, Smith AC, Francis CM (2019) A small-scale response of urban bat activity to tree cover. Urban Ecosyst [Internet]. 22(5):795–805. https://doi.org/10.1007/s11252-019-00846-w

16. Kubista CE, Bruckner A (2015) Importance of urban trees and buildings as daytime roosts for bats. Biologia (Bratisl) [Internet] 70(11):1545–1552. https://doi.org/10.1515/biolog-2015-0179

17. Lim V-C, Clare EL, Littlefair JE, Ramli R, Bhassu S, Wilson J-J (2018) Impact of urbanisation and agriculture on the diet of fruit bats. Urban Ecosyst [Internet]. 21(1):61–70. https://doi.org/10.1007/s11252-017-0700-3

18. Chan AAQ, Aziz SA, Clare EL, Coleman JL (2021) Diet, ecological role and potential ecosystem services of the fruit bat, Cynopterus brachyotis, in a tropical city. Urban Ecosyst [Internet]. 24(2):251–263. https://doi.org/10.1007/s11252-020-01034-x

19. Parris KM, Hazell DL (2005) Biotic effects of climate change in urban environments: the case of the grey-headed flying-fox (Pteropus poliocephalus) in Melbourne, Australia. Biol Conserv [Internet]. 124(2):267–276. Available from: https://www.sciencedirect.com/science/article/pii/S0006320705000625

20. Rollinson DP, Coleman JC, Downs CT (2013) Seasonal differences in foraging dynamics, habitat use and home range size of Wahlberg's epauletted fruit bat in an urban environment. African Zool [Internet] 48(2):340–350. https://doi.org/10.3377/004.048.0218

21. Jara-Servín AM, Saldaña-Vázquez RA, Schondube JE (2017) Nutrient availability predicts frugivorous bat abundance in an urban environment. Mammalia [Internet] 81(4):367–374. https://doi.org/10.1515/mammalia-2015-0039

22. Knop E (2016) Biotic homogenization of three insect groups due to urbanization. Glob Chang Biol [Internet] 22(1):228–236. https://doi.org/10.1111/gcb.13091

23. Threlfall CG, Law B, Banks PB (2012) Influence of Landscape Structure and Human Modifications on Insect Biomass and Bat Foraging Activity in an Urban Landscape. PLoS One [Internet]. 7(6):e38800. https://doi.org/10.1371/journal.pone.0038800

24. Threlfall CG, Williams NSG, Hahs AK, Livesley SJ (2016) Approaches to urban vegetation management and the impacts on urban bird and bat assemblages. Landsc Urban Plan [Internet] 153:28–39. Available from: https://www.sciencedirect.com/science/article/pii/S0169204616300500

25. MacIvor JS, Lundholm J (2011) Insect species composition and diversity on intensive green roofs and adjacent level-ground habitats. Urban Ecosyst [Internet]. 14(2):225–241. https://doi.org/10.1007/s11252-010-0149-0

26. Parkins KL, Clark JA (2015) Green roofs provide habitat for urban bats. Glob Ecol Conserv [Internet]. 4:349–357. Available from: https://www.sciencedirect.com/science/article/pii/S2351989415000840

27. Pearce H, Walters CL (2012) Do green roofs provide habitat for bats in urban areas? Acta Chiropterologica [Internet] 14(2):469–478. https://doi.org/10.3161/150811012X661774

28. Gutierrez E d A, Pessoa VF, Aguiar LMS, Pessoa DMA (2014) Effect of light intensity on food detection in captive great fruit-eating bats, Artibeus lituratus (Chiroptera: Phyllostomidae). Behav Processes [Internet]. 109:64–69. Available from: https://www.sciencedirect.com/science/article/pii/S0376635714001715

29. Haddock JK, Threlfall CG, Law B, Hochuli DF (2019) Light pollution at the urban forest edge negatively impacts insectivorous bats. Biol Conserv [Internet]. 236:17–28. Available from: https://www.sciencedirect.com/science/article/pii/S0006320718313685

30. Diniz UM, Lima SA, Machado ICS (2019) Short-distance pollen dispersal by bats in an urban setting: monitoring the movement of a vertebrate pollinator through fluorescent dyes. Urban Ecosyst [Internet]. 22(2):281–291. https://doi.org/10.1007/s11252-019-0825-7

31. Straka TM, Wolf M, Gras P, Buchholz S, Voigt CC (2019) Tree cover mediates the effect of artificial light on urban bats. Front Ecol Evol [Internet]. 7. Available from: https://www.frontiersin.org/article/10.3389/fevo.2019.00091

32. Voigt CC, Scholl JM, Bauer J, Teige T, Yovel Y, Kramer-Schadt S et al (2020) Movement responses of common noctule bats to the illuminated urban landscape. Landsc Ecol [Internet] 35(1):189–201. https://doi.org/10.1007/s10980-019-00942-4

33. Jung K, Kalko EKV (2011) Adaptability and vulnerability of high flying Neotropical aerial insectivorous bats to urbanization. Divers Distrib [Internet] 17(2):262–274. https://doi.org/10.1111/j.1472-4642.2010.00738.x

34. Domer A, Korine C, Slack M, Rojas I, Mathieu D, Mayo A et al (2021) Adverse effects of noise pollution on foraging and drinking behaviour of insectivorous desert bats. Mamm Biol [Internet] 101(4):497–501. https://doi.org/10.1007/s42991-021-00101-w

35. Li H, Crihfield C, Feng Y, Gaje G, Guzman E, Heckman T et al (2020) The weekend effect on urban bat activity suggests fine scale human-induced bat movements. Animals 10:1636. https://doi.org/10.3390/ani10091636

36. Bunkley JP, McClure CJW, Kleist NJ, Francis CD, Barber JR (2015) Anthropogenic noise alters bat activity levels and echolocation calls. Glob Ecol Conserv [Internet] 3:62–71. Available from: https://www.sciencedirect.com/science/article/pii/S235198941400064X

37. Lehrer EW, Gallo T, Fidino M, Kilgour RJ, Wolff PJ, Magle SB (2021) Urban bat occupancy is highly influenced by noise and the location of water: Considerations for nature-based urban planning. Landsc Urban Plan [Internet]. 210:104063. Available from: https://www.sciencedirect.com/science/article/pii/S0169204621000268

38. Everette AL, O'Shea TJ, Ellison LE, Stone LA, McCance JL (2001) Bat use of a high-plains urban wildlife refuge. Wildl Soc Bull [Internet] 29(3):967–973. Available from: http://pubs.er.usgs.gov/publication/1015150

39. Lintott PR, Bunnefeld N, Park KJ (2015) Opportunities for improving the foraging potential of urban waterways for bats. Biol Conserv [Internet]. 191:224–233. Available from: https://www.sciencedirect.com/science/article/pii/S0006320715300057

40. Ancillotto L, Bosso L, Salinas-Ramos VB, Russo D (2019) The importance of ponds for the conservation of bats in urban landscapes. Landsc Urban Plan [Internet]. 190:103607. Available from: https://www.sciencedirect.com/science/article/pii/S0169204618310211
41. Straka TM, Lentini PE, Lumsden LF, Wintle BA, van der Ree R (2016) Urban bat communities are affected by wetland size, quality, and pollution levels. Ecol Evol [Internet]. 6(14):4761–4774. https://doi.org/10.1002/ece3.2224
42. Straka TM, Lentini PE, Lumsden LF, Buchholz S, Wintle BA, van der Ree R (2020) Clean and green urban water bodies benefit nocturnal flying insects and their predators, insectivorous bats. Sustainability 12:2634. https://doi.org/10.3390/su12072634
43. Blakey RV, Law BS, Straka TM, Kingsford RT, Milne DJ (2018) Importance of wetlands to bats on a dry continent: a review and meta-analysis. Hystrix Ital J Mammal [Internet]. 29(1):41–52. https://doi.org/10.4404/hystrix-00037-2017
44. Gili F, Newson SE, Gillings S, Chamberlain DE, Border JA (2020) Bats in urbanising landscapes: habitat selection and recommendations for a sustainable future. Biol Conserv [Internet]. 241:108343. Available from: https://www.sciencedirect.com/science/article/pii/S0006320719310274
45. Agpalo E, Bennett VJ (2019) Improving urban habitats for bats: what makes a bat-friendly residential swimming pool? In: North American Society for Bat Research 49th annual symposium. Kalamazoo, MI, USA, p. 1–2
46. Gorecki V, Rhodes M, Parsons S (2019) Roost selection in concrete culverts by the large-footed myotis (*Myotis macropus*) is limited by the availability of microhabitat. Aust J Zool [Internet] 67(6):281–289. https://doi.org/10.1071/ZO20033
47. Meierhofer MB, Leivers SJ, Fern RR, Wolf LK, Young JH Jr, Pierce BL et al (2019) Structural and environmental predictors of presence and abundance of tri-colored bats in Texas culverts. J Mammal [Internet]. 100(4):1274–1281. https://doi.org/10.1093/jmammal/gyz099
48. Jones PL, Divoll TJ, Dixon MM, Aparicio D, Cohen G, Mueller UG et al (2020) Sensory ecology of the frog-eating bat, Trachops cirrhosus, from DNA metabarcoding and behavior. Behav Ecol [Internet] 31(6):1420–1428. https://doi.org/10.1093/beheco/araa100
49. Halfwerk W, Blaas M, Kramer L, Hijner N, Trillo PA, Bernal XE et al (2019) Adaptive changes in sexual signalling in response to urbanization. Nat Ecol Evol 3(3):374–380
50. Russo D, Cistrone L, Libralato N, Korine C, Jones G, Ancillotto L (2017) Adverse effects of artificial illumination on bat drinking activity. Anim Conserv [Internet]. 20(6):492–501. https://doi.org/10.1111/acv.12340
51. Russo D, Ancillotto L, Cistrone L, Libralato N, Domer A, Cohen S et al (2019) Effects of artificial illumination on drinking bats: a field test in forest and desert habitats. Anim Conserv [Internet] 22(2):124–133. https://doi.org/10.1111/acv.12443
52. Naidoo S, Vosloo D, Schoeman MC (2013) Foraging at wastewater treatment works increases the potential for metal accumulation in an urban adapter, the banana bat (Neoromicia nana). African Zool [Internet]. 48(1):39–55. https://doi.org/10.1080/15627020.2013.11407567
53. Kalcounis-Rueppell MC, Payne VH, Huff SR, Boyko AL (2007) Effects of wastewater treatment plant effluent on bat foraging ecology in an urban stream system. Biol Conserv [Internet] 138(1):120–130. Available from: https://www.sciencedirect.com/science/article/pii/S0006320707001619
54. Threlfall C, Law B, Penman T, Banks PB (2011) Ecological processes in urban landscapes: mechanisms influencing the distribution and activity of insectivorous bats. Ecography (Cop) [Internet] 34(5):814–826. https://doi.org/10.1111/j.1600-0587.2010.06939.x
55. Fahrig L, Baudry J, Brotons L, Burel FG, Crist TO, Fuller RJ et al (2011) Functional landscape heterogeneity and animal biodiversity in agricultural landscapes. Ecol Lett [Internet] 14(2):101–112. https://doi.org/10.1111/j.1461-0248.2010.01559.x
56. Lindenmayer D, Hobbs RJ, Montague-Drake R, Alexandra J, Bennett A, Burgman M et al (2008) A checklist for ecological management of landscapes for conservation. Ecol Lett [Internet]. 11(1):78–91. https://doi.org/10.1111/j.1461-0248.2007.01114.x

57. Tena E, Fandos G, de Paz Ó, de la Peña R, Tellería JL (2020) Size does matter: passive sampling in urban parks of a regional bat assemblage. Urban Ecosyst [Internet]. 23(2):227–234. https://doi.org/10.1007/s11252-019-00913-2

58. Fabianek F, Gagnon D, Delorme M (2011) Bat distribution and activity in Montréal Island green spaces: responses to multi-scale habitat effects in a densely urbanized area. Écoscience [Internet] 18(1):9–17. https://doi.org/10.2980/18-1-3373

59. Dixon MD (2012) Relationship between land cover and insectivorous bat activity in an urban landscape. Urban Ecosyst [Internet] 15(3):683–695. https://doi.org/10.1007/s11252-011-0219-y

60. Coleman JL, Barclay RMR (2012) Urbanization and the abundance and diversity of Prairie bats. Urban Ecosyst [Internet]. 15(1):87–102. https://doi.org/10.1007/s11252-011-0181-8

61. Caryl FM, Lumsden LF, van der Ree R, Wintle BA (2016) Functional responses of insectivorous bats to increasing housing density support 'land-sparing' rather than 'land-sharing' urban growth strategies. J Appl Ecol [Internet] 53(1):191–201. https://doi.org/10.1111/1365-2664.12549

62. Parker KA, Springall BT, Garshong RA, Malachi AN, Dorn LE, Costa-Terryll A et al (2019) Rapid increases in bat activity and diversity after wetland construction in an urban ecosystem. Wetlands [Internet] 39(4):717–727. https://doi.org/10.1007/s13157-018-1115-5

63. BASHAM R, LAW B, BANKS P (2011) Microbats in a 'leafy' urban landscape: are they persisting, and what factors influence their presence? Austral Ecol [Internet] 36(6):663–678. https://doi.org/10.1111/j.1442-9993.2010.02202.x

64. Van Helden BE, Close PG, Stewart BA, Speldewinde PC, Comer SJ (2020) An underrated habitat: Residential gardens support similar mammal assemblages to urban remnant vegetation. Biol Conserv [Internet]. 250:108760. Available from: https://www.sciencedirect.com/science/article/pii/S0006320720308181

65. Fuller RA, Gaston KJ (2009) The scaling of green space coverage in European cities. Biol Lett [Internet] 5(3):352–355. https://doi.org/10.1098/rsbl.2009.0010

66. Nystrom GS, Bennett VJ (2019) The importance of residential swimming pools as an urban water source for bats. J Mammal [Internet]. 100(2):394–400. https://doi.org/10.1093/jmammal/gyz020

67. Korine C, Adams AM, Shamir U, Gross A (2015) Effect of water quality on species richness and activity of desert-dwelling bats. Mamm Biol [Internet]. 80(3):185–190. Available from: https://www.sciencedirect.com/science/article/pii/S161650471500035X

Chapter 9
Assessing the Effects of Urbanisation on Bats in Recife Area, Atlantic Forest of Brazil

Enrico Bernard, Laura Thomázia de Lucena Damasceno, Alini Vasconcelos Cavalcanti de Frias, and Frederico Hintze

Abstract The metropolitan area of Recife has 4.1 million inhabitants and is in one of the most deforested areas of the Brazilian Atlantic Forest. Here, few forest fragments remain in a large urban matrix. How do bats interact with this complex and challenging landscape? Focusing on insectivores, we used bioacoustics to (1) assess which bats use green spaces in Recife, (2) evaluate the effect of intense artificial light at night (ALAN) on bat activity, and (3) compare molossids' activity and behaviour in urban and non-urban areas. Although with fewer species when compared to forested areas, we show that several insectivores can persist in that urban matrix. However, these species make a heterogeneous use of the landscape: green spaces were hotspots for bat activity with nearly 2.4 times more activity than non-green spaces. We also identified that ALAN from soccer (football) stadiums influences not only the total activity of insectivores but also their temporal activity patterns. However, this influence was stadium-specific with no pattern common to all stadia, pointing to a more complex interaction between species and ALAN. Finally, we show that urbanisation influences the activity and behaviour of urban molossids. We found a significant decrease in total activity in the urban environment but found that urban molossids are active earlier in the night compared to non-urban molossids. Our studies provide evidence of some of the effects of urbanisation on insectivorous bats in a large city among the remains of the Atlantic Forest in Northeastern Brazil, and we highlight the need for more local research and conservation action.

E. Bernard (✉) · L. T. de Lucena Damasceno · A. V. C. de Frias
Laboratório de Ciência Aplicada à Conservação da Biodiversidade, Departamento de Zoologia, Centro de Biociências, Universidade Federal de Pernambuco, Recife, PE, Brazil
e-mail: enrico.bernard@ufpe.br

F. Hintze
Laboratório de Ciência Aplicada à Conservação da Biodiversidade, Departamento de Zoologia, Centro de Biociências, Universidade Federal de Pernambuco, Recife, PE, Brazil

Programa de Pós-Graduação em Biologia Animal, Universidade Federal de Pernambuco, Recife, PE, Brazil

Keywords Artificial light at night (ALAN) · Bioacoustics · Green urban remnants
· Urban fauna

1 Introduction

Urbanisation is transforming the terrestrial environment markedly [1, 2].
Deforestation, habitat degradation, and fragmentation due to urbanisation are affecting thousands of species. Urban expansion can be particularly harmful when it
occurs in areas rich in biodiversity (i.e. biodiversity hotspots - locations with a high
level of biodiversity and endemism that suffer intense degradation and threat to their
diversity) [3]. Extensive tropical regions around the world are impacted by poorly
planned expansion of their urban centres, amplifying the effects of habitat loss and
degradation and threatening species [4].

Perhaps one of the most extreme examples of the impact that urbanisation can
have on a tropical hotspot is the Atlantic Forest of Brazil, where biodiversity is
becoming critically threatened by extensive human occupation. The Atlantic Forest
harbours about 60% of all endangered species in Brazil [5], but about 70% of the
Brazilian population currently resides in the Atlantic Forest domains [6], and less
than 30% of the original forest remains. This forest mostly consists of second-
growth forests or areas affected by edge effects [5, 7].

Although occupation predates the colonisation of Brazil (i.e. more than 500 years
ago), the process of urban densification in the Brazilian Atlantic Forest is relatively
recent, having intensified around the 1970s, when rural populations moved to the
cities. This population growth in cities was rapid and poorly planned. This brought
about a new and heterogeneous landscape: the green spaces that remain within cities
are small parks (frequently <10 hectares), plazas, gardens, and a few private or public reserves of arboreal environment. As in many parts of the world, urban green
spaces of the Brazilian Atlantic Forest are small, isolated, and distant from each
other, forming green islands within the urban matrix [8].

Urbanisation can, however, generate different responses on different wildlife
groups [9–11]. Some wildlife may be highly disturbed by anthropogenic modifications and cannot persist in cities, while others may tolerate or obtain benefits from
anthropogenic environments and human interactions within them (i.e. synanthropic
species). The impact of urbanisation on a species is dependent on its traits [8, 12,
13], and different species of bats present different responses to urbanisation (see
[14]). Some bat families and species are highly sensitive to habitat change, and
urbanisation can strongly impact the structure of local communities, causing a

decrease in richness and diversity when compared to the original pre-urbanisation habitats. Structures such as bridges and buildings may provide roosts for some species, helping them to adapt to the urban environment [15]. Conversely, species specialised in using natural roosts are unlikely to persist in cities if these roosts are limited [16]. Artificial lights attract insects and may improve prey availability for light-tolerant species [17, 18]. In contrast, some species avoid illuminated areas because they are sensitive to artificial light at night (ALAN), and therefore cannot make use of insect clusters around lights [15, 19].

It can be quite challenging to understand the responses of bats – a speciose and ecologically diverse group – to urbanisation in biodiversity hotspots. In Brazil, at least 84 bat species have been recorded in cities [20], representing about 47% of the country's 181 described bats [21], and nearly 100 species occur in the entire Atlantic Forest [22]. However, most of these records are sporadic or from studies of bats and disease transmission. As such, there is a clear and pressing need for more research on these bats, given the projected losses of natural forest with future urban expansion in the region.

In this chapter, we present three studies of the effects of urban modification on bats in the area of Recife, the capital of Pernambuco State, in the Northeastern Brazilian Atlantic Forest. We focused on insectivorous species using bioacoustics to (1) assess which species use urban green spaces, (2) evaluate the effect of artificial light on bat activity, and (3) compare the activity and behaviour of urban- and non-urban-dwelling species.

Recife (8°03′46″S, 34°52′14″W) is Brazil's third most densely populated metropolitan area. It consists of 15 municipalities distributed across ca. 3200 km², with ~4.1 million human inhabitants [23]. The climate is tropical, with average annual temperature ranging from 24 °C to 26 °C and 1800 and 2000 mm of rain annually (https://portal.inmet.gov.br).

The history of urbanisation in the region goes back 500 years. Indigenous villages historically existed in the Atlantic Forest, but the first urban settlements were established by the Portuguese in the 1530s. Since this colonisation, the region's economy has been heavily dependent on extraction of natural resources and sugarcane farming. Over time, these activities have driven deforestation and forest degradation, leaving behind small, isolated, and poor-quality habitat patches.

The 15 municipalities are totally inserted in the Atlantic Forest biome and have ~640 km² of natural remnants or ~19.5% of their total area, varying from 2 to 108 km² of remnants per municipality (4.8–61.4% of their areas). An inventory of bat species has never been conducted in Recife's metropolitan region, but small-scale short-term inventories based on mist netting have resulted in records of 12–26 species per site [24], and a tentative list for the area presents at least 59 species in 37 genus and 5 families.

2 Bat Species Richness and Activity in Green Spaces and Urban Areas in Recife

To examine how green spaces affect insectivorous bats, we measured bat activity at sampling sites across the region. We used satellite images to select five study sites. Each site had a green space measuring between 15,239 and 139,457 m² and a paired urban area 200 m away. Between September 2012 and January 2013 (five sampling nights per site), we simultaneously monitored the green space and the urban area for 2 h per night. We recorded bat activity on full-spectrum bat detectors, which we set up in locations expected to have low human traffic.

We used CallViewer18 [25] to view and analyse files and extract key call parameters (i.e. duration, minimum and maximum frequencies, frequency with maximum energy, and call intensity), which we used to assign search-phase calls to sonotypes and families, based on published values [26]. For a detailed description of the methodology adopted, see [27].

We analysed 500 files and 1500 min of recordings. We recorded a total of 16 sonotypes (16 in green spaces, 8 in urban areas) belonging to 5 families (Emballonuridae, Molossidae, Noctilionidae, Phyllostomidae, Vespertilionidae) – all but the Phyllostomids are strict insectivores. Compared to the paired urban areas, green spaces had more bat species and activity (student t: $t = 2.53$; $P = 0.02$), but not more feeding buzzes ($t = 1.81$; $P = 0.08$) or social calls ($t = -1.56$; $P = 0.13$) (Fig. 9.1). Vespertilionids were more active in green spaces (Chi-square residual test $R = 6.27$), while molossids were more active in urban areas ($R = 4.71$).

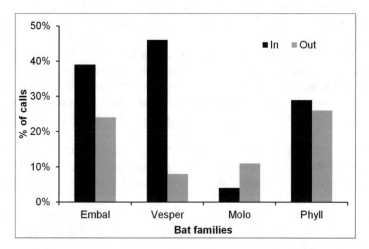

Fig. 9.1 Activity of four different bat families recorded in green urban remnants (GR) and outside them (NR) in the metropolitan area of Recife, Northeastern Brazil. Embal = Emballonuridae; Vesper = Vespertilionidae; Molo = Molossidae; Phyll = Phyllostomidae. Noctilionidae was recorded but accounted for <1% of the calls and was not represented. Activity is expressed based on the % of echolocation calls in the analysed files, and one file could contain calls from more than one family. Therefore, the sum of the percentages could be higher than 100%

3   Impact of Artificial Illumination on Bat Activity

Urbanisation increases light pollution, but the ecological impacts already extend well beyond the urban fringe and increasingly affect species in natural habitats, including protected areas [28]. These impacts occur because artificial light can interfere with species' ability to access resources; some prey species may avoid bright areas due to a higher risk of predation, while other predator species may prefer those areas because they may attract more prey – like insects around a streetlight – making easier to locate them [29, 30]. ALAN may modify light-mediated endocrine regulations, causing shifts in reproduction patterns [11, 36], and may also disrupt the spatial orientation of nocturnal species, by removing dark areas or creating artificial shadows and reflexes [30, 31]. Impacts on species appear to differ among taxa but may ultimately have cascading ecosystem effects that translate into biodiversity loss [31–33].

Patterns of artificial illumination in cities are complex and heterogeneous with spatial and temporal variation in the intensity of light. For example, sports stadia concentrate intense lighting on game nights but are dark when not in use – as such, they are interesting locations to study ecological impacts of lights [34]. Bats are suitable study subjects, given that they are strictly nocturnal and may use vision to find food and detect predators ([30] and others therein).

Given knowledge that many phototactic insects are attracted to certain artificial lights [15, 30, 34–37], we tested the prediction that bat activity at football (soccer) stadia is greater on game nights, when lights are on, than on non-game nights, when they are off, and asked whether the lights otherwise influence bat behaviour.

We selected three stadia, all fitted with 2000-W mercury vapour lights. The stadium Ilha do Retiro is 16,515 m², can hold 32,983 people, and has 96 reflectors. It is located next to a river and has swimming pools, tennis courts, shops, car park, and smaller soccer fields, as well as billboards that are lit every night. It is mainly surrounded by residences and shops (67% in a 500-m buffer). Arruda Stadium, similarly surrounded by residential and commercial buildings (98% in a 500-m buffer), is 30,000 m², seats 60,044 spectators, and has 128 reflectors. Finally, Arena Pernambuco is 24,000 m², with a capacity of 44,300 people and 360 reflectors, and is surrounded by forest fragments (51% vegetation and no buildings in a 500-m buffer).

Between April and September 2018, we acoustically monitored bats on a total of 11 game nights and 41 non-game nights. We worked four times at Ilha do Retiro and three times at Arruda – at each stadium, we recorded bats on the eve of the game, on game night, and on the three subsequent nights (i.e. 1 off, 1 lit, 3 off). We obtained four samples at Arena Pernambuco using the same protocol but instead recorded on game night and on four subsequent nights (i.e. 1 lit, 4 off). We used full-spectrum bat detectors, placed under one reflector in the centre of the stadiums, and recorded from 1 h before sunset to 1 h after sunrise. We used Kaleidoscope (version 4.5.5, Wildlife Acoustics, Maynard, MA, USA) to process and analyse spectrograms. We

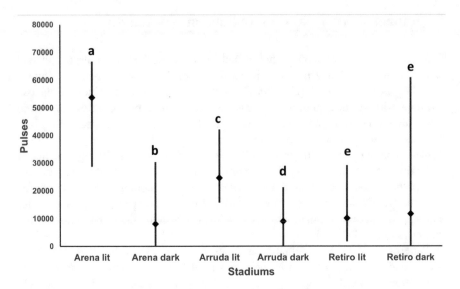

Fig. 9.2 Bat echolocation pulses recorded in three football stadiums (Arena Pernambuco, Arruda, and Ilha do Retiro) in the metropolitan area of Recife, Northeastern Brazil, between April and September 2018. Bat pulses were recorded from 16:00 h to 06:00 h along nights with reflectors on (lit) and off (dark). Diamonds indicate the average number of pulses and bars the interval variation. Different letters represent statistically significant differences within site pairs

used the number of echolocation calls as a proxy for bat activity and Mann-Whitney to test for differences between game and non-game nights.

We observed more bat activity on game nights than on non-game nights at Arena Pernambuco ($P = 0.002$) and Arruda ($P = 0.0167$), but not at Ilha do Retiro ($P = 0.44$) (Fig. 9.2). We also observed shifts in temporal activity patterns, but not consistent ones. At Arena Pernambuco, for example, activity patterns were trimodal on non-game nights (peaks at ca. 17:30 h, 00:30 h, and 4:30 h) and bimodal on game nights (peaks at ca. 20:00 h and 04:30 h) (Fig. 9.3). At Ilha do Retiro, bats exhibited a bimodal pattern on non-game nights (peaks at ca. 17:30 h and 05:00 h) and a tri-modal pattern on game nights (peaks at ca. 17:00 h, 23:00 h, and 4:30 h). At Arruda, bat activity was unimodal on non-game nights (peak at ca. 17:30 h) and trimodal on game nights (peaks at ca. 17:30 h, 21:30 h, and 4:30 h).

4   Molossid Activity and Behaviour in Urban vs. Non-urban Areas

The family Molossidae accounts for almost one third of the country's total diversity of insectivorous bats [21] and is one of the main insectivorous bat families in cities [14, 18, 27, 38]. Several authors (e.g. [16, 18, 39, 40]) argue that many molossids

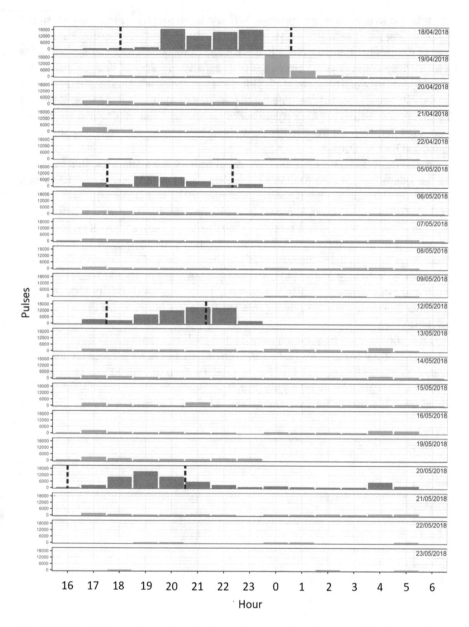

Fig. 9.3 Bat activity based on the recording of echolocation calls along 19 nights at Arena Pernambuco Stadium, in Recife, Northeastern Brazil. Echolocation calls were recorded on nights with reflectors on (red) and off (light blue), and dashed vertical lines mark the intervals where the lights were on

are adapted to the urban environment. Molossids have a fast flight style well suited to open areas and with few obstacles [41]. Furthermore, their roosting habits (i.e. use of fissures or crevices), allow them to take advantage of the many opportunities provided by anthropogenic structures (e.g. expansion joints of bridges and buildings, spaces between tiles, etc.) [15, 39]. However, molossids are still poorly studied in the Neotropics, because mist nets – the traditional method of capturing bats – are relatively ineffective for high-flying aerial insectivores like molossids. Here, bioacoustics can help fill knowledge gaps about habitat use and activity patterns of molossid bats in cities.

In our third study, we monitored molossid activity in an urban site in Recife and in the non-urban, 562-ha Saltinho Biological Reserve (8°43′57″S, 35°10′25″W), also in the Atlantic Forest. Between December 2017 and December 2018, we set full-spectrum bat detectors to record in open areas for 4 h (starting 1 h before sunset) on calm nights without rain. We visualised spectrograms in Raven Pro 1.5 (Cornell Lab of Ornithology, USA) and assigned calls to species by comparing recordings with published literature [26]. We used bat activity as a proxy for relative abundance, following Miller [42], and analysed activity in 10 min intervals within each night of sampling. Finally, we compared total molossid activity between both environments using student t-tests (after verifying that our data met the assumption of normality).

Of 6,866 bat records obtained over 24 h in each environment, 47.8% were from molossids. Total bat activity in the forest environment (789.7 ± 235.2 records) was significantly higher ($t = 3.4102$; $p = 0.0270$) than in the urban environment (324.3 ± 22.9 records) on all nights (Fig. 9.4). However, whereas molossids clearly dominated the urban bat assemblage (68–82% of records per night), they only

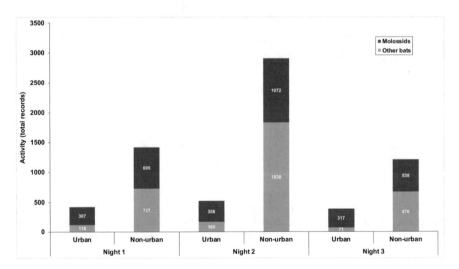

Fig. 9.4 Activity of molossid and other insectivorous bats in non-urban (Saltinho Reserve, a 562-hectare forest remnant) and urban (Recife) areas, in the northeastern Atlantic Forest of Brazil. Data based on 24 h of recordings in each habitat along three simultaneous nights between December 2017 and December 2018

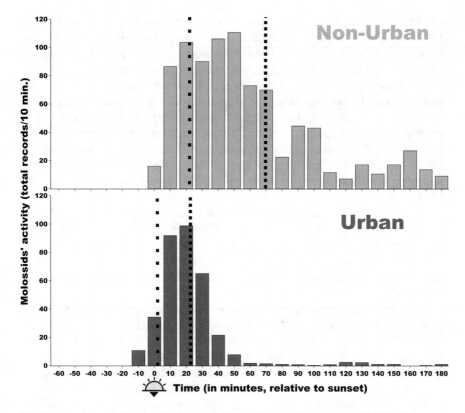

Fig. 9.5 Average activity of molossids in relation to the sunset in non-urban (Saltinho Reserve, a 562-hectare forest remnant) and urban (Recife) areas, in the northeastern Atlantic Forest of Brazil. The *x* axis represents 10 min time blocks and 0 marks the sunset. The first dashed vertical lines mark 25% of the recorded activity, and second dashed vertical lines mark 75% of the recorded activity. Data based on 24 h of recordings in each habitat along three simultaneous nights between December 2017 and December 2018

accounted for 37–49% of the total bat activity in the reserve. Furthermore, urban molossids started about 10 min before sunset and were mainly active in the early part of the night, with 75% of the total activity in the first 25 min after sunset (Fig. 9.5). In contrast, molossid activity in the reserve started later and was more homogeneous, with 75% occurring in the first 90 min after sunset.

5   Discussion and Conclusions

Our three studies offer glimpses of some of the effects of urbanisation on insectivorous bats in a large city within Brazil's Atlantic Forest region. First, although the urban landscape is more simplified than natural forest, several bat species (represented by different sonotypes) persist in the city of Recife. However, our results also suggest that those bats use this urban landscape heterogeneously. As prior studies [14] have reported, a combination of ecological and behavioural adaptability and opportunism determines whether a bat species can exploit the diverse anthropogenic resources in the urban matrix and ultimately tolerate urbanisation. The concept of "urban adapters" and "urban avoiders" is appropriate and helps to elucidate which species do well in and which ones are extirpated from urban areas [14, 18]. In our first study, we recorded 16 sonotypes belonging to 5 bat families in Recife – presumably, these are all at least somewhat urban "adapted" or tolerant. Yet, green spaces matter: they supported twice as many species and nearly 2.4 times more bat activity compared to paired urbanised sites. In fact, green spaces were bat hotspots, suggesting that urban green spaces are top priorities for bat conservation (see also [27, 43, 44]).

Our second study revealed that the intense lighting of sports stadia can influence bats, but contextually in that bat activity increased when lights were on, but only at two stadia. Furthermore, stadium lights appear to influence temporal activity patterns in a contextual way, which suggests that impacts of artificial light on bats reflect complex interactions between multiple variables, including species-specific responses.

All terrestrial (above-ground) life evolved in environments where natural light regimes vary daily and seasonally in predictable ways, and nocturnal animals are particularly responsive to alteration of these regimes [45]. Indeed, light is the most important environmental factor that regulates the activity of bats [46], and artificial light can alter interactions between species that use very similar resources in a lighting gradient [26]. Thus, bats are expected to be affected by artificial light (e.g. [30, 47–49]), but two key points must be considered. First, impacts depend on light spectra and intensity. Second, bat responses are species-specific. Some insectivores may benefit by exploiting swarms of phototactic insects at lights, while others avoid light. In fact, studies ([30] and others therein) document altered feeding patterns of several bat species resulting from light pollution and that these changes ultimately act as a species filter and tend to homogenise urban bat faunas. Although we used sonotype as a proxy for species, identifying acoustic recordings of urban bats to species is useful to obtain a better understanding of which ones are positively or negatively affected by urbanisation. Doing so in Neotropical cities becomes increasingly feasible as the number and quality of published reference call libraries increase (e.g. [26]).

Unfortunately, urban light pollution is not considered a conservation threat (or even a pressing problem) in many lower- to middle-income countries. In fact, an increase in lit areas is frequently seen as necessary and beneficial due to the perception (justified or not; [50]) that lighting reduces crime and improves traffic safety. However, with mounting evidence that light pollution affects not only urban flora and fauna but also human health [29, 31, 45, 51], it is worth reducing and mitigating urban light pollution. Bats could benefit from actions to do so, especially by adopting bat-friendly lightning [52].

Our third study indicated that urbanisation influences the activity and behaviour patterns of molossids. Relative to their counterparts in the native Atlantic Forest, urban bats appeared to exhibit reduced activity that was also concentrated in a shorter interval near the beginning of the night. The observed early foraging activity by urban molossids may result from several causes, one of which may reflect responses to possible changes in prey availability. The lower abundance and diversity of insects in urban areas can intensify competition among bats – to avoid competitors, they may anticipate the beginning of foraging [53]. Previous studies of insectivorous bats, and especially molossids, have found that for them, optimal foraging implies timing their activities to those of their prey and predators [18, 36, 37, 54]. For example, artificial lights, by attracting swarms of phototactic insects, may in turn attract hungry molossids, which spend less time searching for prey and thus benefit from reduced energy expenditure and greater hunting efficiency [17, 39, 47]. For molossids, this is possible thanks to their fast flight style, which mitigates the risk of exposure to and capture by visual predators. In addition, molossids may benefit from reduced competition with slow-flying bats, which tend to avoid brightly lit areas where they are vulnerable to predation.

Alternatively, urbanisation could affect the foraging behaviour of molossids, forcing them to move to less urbanised, peripheral areas later in the night, because the biomass of moths, which are important in their diet, is reduced in bright environments [16, 17, 39, 55, 56]. This change in the diversity and abundance of potential prey in illuminated areas could explain why the activity of urban molossids declines rapidly after the beginning of the night. Both hypotheses should be tested.

Despite their shortcomings (e.g. small sample size, short duration, focus on single variables), our three studies offer useful baselines and examples of issues to address when investigating interactions between Neotropical bats and urban environments. Such investigations are worthwhile not only to determine the drivers of bats' responses but also to identify strategies to mitigate the negative impacts of urbanisation on insectivorous bats, whose ecological roles are vital to ecosystem integrity and the maintenance of biodiversity.

Literature Cited

1. Seto KC, Fragkias M, Güneralp B, Reilly MK (2011) A meta-analysis of global urban land expansion. PLoS One 6(8):e23777
2. Shochat E, Warren PS, Faeth SH et al (2006) From patterns to emerging processes in mechanistic urban ecology. Trends Ecol Evol 21:186–191
3. Seto KC, Güneralp B, Hutyra LR (2012) Global forecasts of urban expansion to 2030 and direct impacts on biodiversity and carbon pools. PNAS 109:16083–16088
4. McDonnell M, Hahs A (2013) The future of urban biodiversity research: moving beyond the 'low-hanging fruit'. Urban Ecosyst 16:397–409
5. Rezende CL, Scarano FR, Assad ED et al (2018) From hotspot to hopespot: an opportunity for the Brazilian Atlantic Forest. Perspect Ecol Conserv 16:208–214
6. Tabarelli M, Aguiar AV, Cesar MR et al (2010) Biodiversity conservation in the Atlantic forest: lessons from aging human-modified landscapes. Biol Conserv 143:2328–2340

7. Ribeiro MC, Metzger JP, Martensen AC et al (2009) The Brazilian Atlantic Forest: how much is left, and how is the remaining forest distributed? Implications for conservation. Biol Conserv 142:1141–1153

8. McKinney ML (2006) Urbanization as a major cause of biotic homogenization. Biol Conserv 127:247–260

9. Aronson MFJ, La Sorte FA, Nilon CH et al (2014) A global analysis of the impacts of urbanization on bird and plant diversity reveals key anthropogenic drivers. Proc R Soc B Biol Sci 281(1780):20133330

10. Chamberlain DE, Cannon AR, Toms MP, Leech DI, Hatchwell BJ, Gaston KJ (2009) Avian productivity in urban landscapes: a review and meta-analysis. Ibis 151:1–18

11. Jokimäki J, Kaisanlahti-Jokimäki M-L, Suhonen J et al (2011) Merging wildlife community ecology with animal behavioral ecology for a better urban landscape planning. Landsc Urban Plan 100:383–385

12. McKinney ML (2002) Urbanization, biodiversity, and conservation. Bioscience 52:883–890

13. Threlfall CG, Law B, Penman T, Banks PB (2011) Ecological processes in urban landscapes: mechanisms influencing the distribution and activity of insectivorous bats. Ecography 34:814–826

14. Jung K, Threlfall CG (2016) Urbanization and its effects on bats – a global meta-analysis. In: Voigt C, Kingston T (eds) Bats in the anthropocene: conservation of bats in a changing world. Springer, Cham, pp 13–33

15. Russo D, Ancillotto L (2015) Sensitivity of bats to urbanization: a review. Mamm Biol 80:205–212

16. Threlfall CG, Law B, Banks PB (2012) Influence of landscape structure and human modifications on insect biomass and bat foraging activity in an urban landscape. PLoS One 7(6):e38800

17. Haddock JK, Threlfall CG, Law B, Hochuli DF (2019) Responses of insectivorous bats and nocturnal insects to local changes in street light technology. Austral Ecol 44:1052–1064

18. Jung K, Kalko EKV (2010) Where forest meets urbanization: foraging plasticity of aerial insectivorous bats in an anthropogenically altered environment. J Mammal 91:144–153

19. Threlfall CG, Law B, Banks PB (2013) The urban matrix and artificial light restricts the nightly ranging behaviour of Gould's long-eared bat (*Nyctophilus gouldi*). Austral Ecol 38:921–930

20. Nunes H, Rocha FL, Cordeiro-Estrela P (2017) Bats in urban areas of Brazil: roosts, food resources and parasites in disturbed environments. Urban Ecosyst 20:953–969

21. Garbino GST, Gregorin R, Lima IP et al (2020) Updated checklist of Brazilian bats: versão 2020. Comitê da Lista de Morcegos do Brasil - CLMB, Sociedade Brasileira para o Estudo de Quirópteros (SBEQ). https://www.sbeq.net/lista-de-especies

22. Muylaert RL, Stevens RD, Esbérard CEL et al (2017) ATLANTIC BATS: a data set of bat communities from the Atlantic Forests of South America. Ecology 98:3227–3227

23. Instituto Brasileiro de Geografia e Estatística IBGE (2021) Cidades: Recife. https://www.ibge.gov.br/cidades-e-estados/pe/recife.html

24. Leal ESB, Guerra Filho DQ, Ramalho DF, Silva JM, Bandeira RS, Silva LAM, Oliveira MAB (2019) Bat Fauna (Chiroptera) in an urban environment in the Atlantic Forest, northeastern Brazil. Neotrop Biol Conserv 14:55–82

25. Skowronski MD, Fenton MB (2008) Model-based automated detection of echolocation calls using the link detector. J Acoust Soc Am 124:328–336

26. Arias-Aguilar A, Hintze F, Aguiar LMS, Rufray V, Bernard E, Pereira MJR (2018) Who's calling? Acoustic identification of Brazilian bats. Mamm Res 63:231–253

27. Araújo MLVS, Bernard E (2016) Green remnants are hotspots for bat activity in a large Brazilian urban area. Urban Ecosyst 19:287–296

28. Mu H, Li X, Du X, Huang J, Su W, Hu T, Wen Y, Yin P, Han Y, Xue F (2021) Evaluation of light pollution in global protected areas from 1992 to 2018. Remote Sens 13:1849

29. Gaston KJ, Davies TW, Bennie J et al (2012) Reducing the ecological consequences of night-time light pollution: options and developments. J Appl Ecol 49:1256–1266

30. Rowse EG, Lewanzik D, Stone EL, Harris S, Jones G (2016) Dark matters: the effects of arti-ficial lighting on bats. In: Voigt C, Kingston T (eds) Bats in the anthropocene: conservation of bats in a changing world. Springer, Cham, pp 187–213
31. Falchi F, Cinzano P, Elvidge CD et al (2011) Limiting the impact of light pollution on human health, environment and stellar visibility. J Environ Manag 92:2714–2722
32. Davies TW, Bennie J, Inger R et al (2013) Artificial light pollution: are shifting spectral signa-tures changing the balance of species interactions? Glob Change Biol 19:1417–1423
33. Longcore T, Rich C (2004) Ecological light pollution. Front Ecol Environ 2:191–198
34. Schoeman MC (2016) Light pollution at stadiums favors urban exploiter bats. Anim Conserv 19:120–130
35. Acharya L, Fenton MB (1999) Bat attacks and moth defensive behaviour around street lights. Can J Zool 77:27–33
36. De Jong J, Ahlén I (1991) Factors affecting the distribution pattern of bats in Uppland, Central Sweden. Ecography 14:92–96
37. Jones G, Rydell J (1994) Foraging strategy and predation risk as factors influencing emergence time in echolocating bats. Philos Trans R Soc Lond Ser B Biol Sci 346:445–455
38. Mora EC, Macías S, Vater M et al (2004) Specializations for aerial hawking in the echoloca-tion system of *Molossus molossus* (Molossidae, Chiroptera). J Comp Physiol A 190:561–574
39. Jung K, Kalko EKV (2011) Adaptability and vulnerability of high-flying Neotropical aerial insectivorous bats to urbanization. Divers Distrib 17:262–274
40. Voigt C, Phelps K, Aguirre L et al (2016) Bats and buildings: the conservation of synanthropic bats. In: Voigt CC, Kingston T (eds) Bats in the anthropocene: conservation of bats in a chang-ing world. Springer, pp 427–462
41. Marinello MM, Bernard E (2014) Wing morphology of Neotropical bats: a quantitative and qualitative analysis with implications for habitat use. Can J Zool 92:141–147
42. Miller BW (2001) A method for determining relative activity of free flying bats using a new activity index for acoustic monitoring. Acta Chiropt 3:93–105
43. Basham R, Law B, Banks P (2011) Microbats in a 'leafy' urban landscape: are they persisting, and what factors influence their presence? Austral Ecol 36:663–678
44. Hale JD, Fairbrass AJ, Matthews TJ, Sadler JP (2012) Habitat composition and connectiv-ity predicts bat presence and activity at foraging sites in a large UK conurbation. PLoS One 7(3):e33300
45. Hölker F, Wolter C, Perkin EK, Tockner K (2010) Light pollution as a biodiversity threat. Trends Ecol Evol 25:681–682
46. Halle S, Stenseth NC (2000) Activity patterns in small mammals – an ecological approach. Springer-Verlag, Berlin Heidelberg
47. Stone EL, Harris S, Jones G (2015) Impacts of artificial lighting on bats: a review of challenges and solutions. Mamm Biol 80:213–219
48. Straka T, Greif S, Schultz S et al (2020) The effect of cave illumination on bats. Global Ecol Conserv 21:e00808
49. Voigt CC, Azam C, Dekker J et al (2018) Guidelines for consideration of bats in lighting proj-ects, EUROBATS publication series no. 8. UNEP/EUROBATS Secretariat, Bonn, p 62
50. Steinbach R, Perkins C, Tompson L, Johnson S, Armstrong B, Green J, Grundy C, Wilinson P, Edwards P (2015) The effect of reduced street lighting on road casualties and crime in England and Wales: controlled interrupted time series analysis. J Epidemiol Community Health 69:1118–1124
51. Navara KJ, Nelson RJ (2007) The dark side of light at night: physiological, epidemiological, and ecological consequences. J Pineal Res 43:215–224
52. Institution of Lighting Professionals (2018) Bats and artificial lighting in the UK, Guidance Note 08/18. Institution of Lighting Professionals and Bat Conservation Trust
53. Arlettaz R, Godat S, Meyer H (2000) Competition for food by expanding pipistrelle bat populations (*Pipistrellus pipistrellus*) might contribute to the decline of lesser horseshoe bats (*Rhinolophus hipposideros*). Biol Conserv 93:55–60

54. Taylor RJ, Oneill MG (1988) Summer activity patterns of insectivorous bats and their prey in Tasmania. Wildl Res 15:533–539
55. Bohmann K, Monadjem A, Noer CL, Rasmussen M, Zeale MRK, Clare E, Jones G, Willerslev E, Gilbert MTP (2011) Molecular diet analysis of two African free-tailed bats (Molossidae) using high throughput sequencing. PLoS One 6:e21441
56. Rodríguez-Aguilar G, Orozco-Lugo CL, Vleut I, Vazquez L-B (2017) Influence of urbanization on the occurrence and activity of aerial insectivorous bats. Urban Ecosyst 20:477–488

Part III
How do Bats and Humans Interact in Urban Environments? Human Perceptions, Public Health, and Ecosystem Services of Bats

Chapter 10
Human Dimensions of Bats in the City

Leonardo Ancillotto, Joanna L. Coleman, Anna Maria Gibellini, and Danilo Russo

Abstract Cities are characterised by low amounts of natural habitat, so their human populations, i.e. urbanites, are expected to be poorly connected to and knowledge-able about the natural world, despite high biodiversity levels being increasingly recorded in unconventional, urban habitats. Such disconnection may raise potential conservation issues for wildlife living in urban areas. Bats are common in cities around the world and so too, therefore, are interactions between bats and urbanites. Yet, the elusive habits and peculiar adaptations of bats make these mammals poorly known and even feared, often as a consequence of long-standing, negative cultural framing. Here, we review the available literature on the human dimensions of urban bats. We first present a potential theoretical framework for understanding the drivers of human-bat relationships and how it applies to the available literature on urban bats. Next, we present an array of potential real-life contexts in which human-bat interactions may occur in urban areas worldwide. Such interactions vary in their nature and context and include visits to zoos, rescue, volunteering, occasional encounters, and the installation of bat boxes. Finally, we present a focal case study investigating attitudes towards and knowledge of bats from a tropical city.

Keywords Attitudes · Bat boxes · Cognitive hierarchy theory · Human-wildlife conflict · Urbanites

L. Ancillotto (✉) · D. Russo
Wildlife Research Unit, Dipartimento di Agraria, Università degli Studi di Napoli Federico II, Naples, Italy
e-mail: leonardo.ancillotto@unina.it

J. L. Coleman
Department of Biology, Queens College at the City University of New York, Flushing, NY, USA

A. M. Gibellini
Centro Recupero Animali Selvatici WWF di Valpredina, Bergamo, Italy

1 Introduction

In this so-called urban century, natural habitats are increasingly replaced by urban land uses, and the proportion of the human population residing in cities is growing. The most urbanised areas usually only host remnants of natural habitats and highly modified wildlife communities in comparison to natural areas [1]. Consequently, the maintenance and management of urban biodiversity are widely accepted as fundamental tasks for land managers, particularly in Western Europe and North America, with "green" strategies often implemented to conserve urban biodiversity and manage interactions between urbanites and wildlife (e.g. [2]).

These interactions are diverse in how they are perceived by humans. Some are arguably unequivocally positive, such as when an attic houses a colony of endangered bats, which otherwise lack suitable roost habitat, and the homeowners are happy about it and enjoy watching the bats emerge at dusk but otherwise leave them alone. Some are benign, as when a pedestrian stops momentarily to watch a bat hunting insects at a streetlamp. Some interactions are unequivocally negative: an urbanite has an aggressive encounter with a bat and kills it. But often, deeming an interaction as positive or negative depends on perspective. For example, a group of people crowds around a roosting bat, enjoying taking photos (positive) while disturbing the bat (negative). Thus, these interactions (and perceptions thereof) are complex.

These interactions also have potentially significant implications for urban conservation because conservation threats to bats (like other wildlife) mostly result from human behaviours [3, 4], and willingness to conserve (like other pro-environmental behaviours) is shaped by our experiences in nature ([5, 6] and others therein). However, due to the relative scarcity of natural habitats in cities, urbanites may be less connected to nature than those living in more rural or natural contexts ([5], but see [6, 7]). Consequently, the human-wildlife relationship in cities may be weak and not based on direct experience, despite the active engagement of people with urban green spaces, especially for recreational activities and wellbeing.

Bats are elusive mammals, rarely encountered by people generally, except perhaps when conflicts arise (e.g. when bats occupy spaces in buildings and are thus noticed by people inhabiting the building) [8] or when a grounded bat is found. Nevertheless, bats are the most speciose order of mammals in many urban areas, where they often have large populations [9]. Because bats are ubiquitous in cities, it is fairly common for urbanites to encounter bats – this makes management of human-bat interactions a fundamental challenge of bat conservation. Also, recent media representations of bats as reservoirs of dangerous viral diseases [10] have increased the importance of understanding human behaviours (and drivers thereof) towards bats.

Studying these interactions and behaviours is an endeavour for human dimension research. Here, we provide a review of what is known about human dimension research as it relates to bats and urbanites (including their knowledge, behaviours, and attitudes towards bats), outlining trends and gaps in this field. We first present a potential theoretical framework for understanding the drivers of human-bat relationships and how this applies to the available literature on the topic of urban bats. Next, we present an array of potential real-life contexts in which human-bat interactions may actually occur in urban areas worldwide, with a focal case study from a tropical city.

2   A Brief Theoretical Framework

The discipline of human dimension research, which falls under the umbrella of conservation social science, integrates various social sciences (e.g. sociology, psychology) to understand stakeholders, including urbanites, in wildlife issues [11]. As such, it employs key social-science conceptual frameworks to study human behaviours (actions performed towards a target object) and their drivers. As an illustrative example, let us consider a homeowner who has a colony of bats in their attic and wishes to get rid of them.

One model, Cognitive Hierarchy Theory, aims to disentangle the cognitive mechanisms that ultimately produce a given human behaviour [12]. In this example, it might be hiring an exterminator. Knowledge and behaviours have a hierarchical structure, i.e. forming a so-called inverted pyramid (Fig. 10.1) [13]. At the base are general values and norms, which are stable and usually resist modification (e.g. via education). These underpin values and norms that are much more contextual and ultimately give rise to the great variety of potential behaviours, which, in contrast to the base of the pyramid, may be rapidly changed through experience. Norms and attitudes represent evaluations of target objects and judgements about them in specific contexts. For example, if "bats" is the object, then a person's evaluation of bats should summarise their general attitude towards bats, while changing the object into

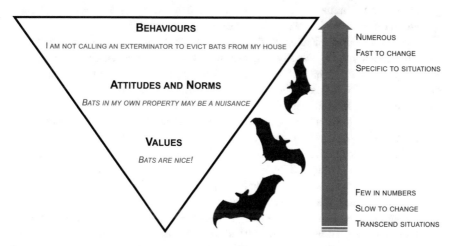

Fig. 10.1 The Cognitive Hierarchy model of human behaviour, with examples of potential evaluations of bats in real-life contexts. (Adapted from Vaske and Donnelly [13])

"bats in your own house" significantly narrows the context and thus will elicit a more circumstance-specific evaluation [14].

Two of the most applied frameworks to pro-environmental behaviours, the theories of Reasoned Action and Planned Behaviour, recognise a person's intent to perform a behaviour (calling an exterminator) as the proximate predictor of whether they ultimately do so ([15] and others therein). This intention has two direct antecedents: first, the person's attitudes about the behaviour (perceived likelihood that hiring an exterminator will effectively get rid of the bats) and, second, their subjective norms about it (belief as to whether certain important others, such as relatives, neighbours, etc., would (1) approve of them hiring an exterminator and (2) hire an exterminator themselves). The Theory of Planned Behaviour further recognises the role of behavioural control – their actual or perceived ability to perform the behaviour (e.g. availability of time and money to hire an exterminator) in modulating the influences of attitudes and norms ([15] and others therein).

Alternate models may distinguish between the cognitive (knowledge and perceptions), normative (involving norms), and affective (emotional) aspects of behaviour (e.g. [16]). Emotions are classifiable by various schemes, including dichotomised between positive (e.g. happiness, surprise) and negative (e.g. sadness, fear; [17] and others therein). Finally, most frameworks acknowledge the importance of other intrinsic behavioural antecedents (e.g. demographics, ethnicity, education, income, personality), though often only as indirectly influential ([15] and others therein).

3 Disentangling Values, Attitudes, and Behaviours

Historically, bat-related human-dimension research, both generally and in urban areas, is relatively scant [18]. Yet, some studies lay the foundations for our understanding of how human behaviours are affected by attitudes and how these may change in different socio-economic contexts that also include urbanites.

A seminal study by Kellert [19] attempted to highlight drivers of negative attitudes towards animals generally, and specifically invertebrates which, like bats, tend to be unpopular [20]. The authors developed and used a questionnaire to identify the cognitive bases for negative behaviours and postulated that the public may perceive these "unloved" animals with fear, antipathy, and aversion, as a result of a generalised fear of the unfamiliar. Where bats are concerned, people may avoid them, selectively focusing on the fact that they may be reservoirs of diseases, especially when the media represents them as such [21]. People may dislike bats because (i) they are morphologically and behaviourally different from people and other species [21], (ii) many species display erratic flight patterns, and (iii) some species "invade" human spaces unexpectedly. They may also dislike bats because they are considered to be negative omens in some folklore, although perhaps mainly in countries where Western cultures predominate, in contrast to many Asia-Pacific cultures which feature positive associations with bats [22]. Nevertheless, local belief systems do not always align with actual practices, i.e. positive cultural framing of bats may also be associated with practices detrimental to bat populations (e.g. unsustainable harvesting; [22]). Following Kellert [23], several authors adopted a Bat Attitude Questionnaire (BAQ) to explore the dimensions of attitudes towards bats [24], disentangling the different roles of myths and knowledge, as well as those of ecologistic, scientistic, and negativistic typologies of human attitudes (see also [25]). Namely, Boso et al. [24] found that gender, educational level, and religiosity are significant drivers of attitude polarisation towards bats. Yet, only a few of these studies actually focused on urbanites or included cities.

A government report from Fort Collins, Colorado [26], directly tackled bat-related attitudes, awareness, and knowledge using a questionnaire administered in 2004. The fact that most respondents, who self-reported living in urban and suburban areas, were also aware of bats' presence and had observed bats regularly (e.g. in their neighbourhoods) supports the statement that despite having lost significant fractions of their original faunas, cities may still offer residents the chance to encounter bats. However, despite being relatively familiar with bats (i.e. given their direct observations), these respondents were not knowledgeable about bats. For example, people were not well-acquainted with ecosystem services provided by bats and were likely to report false information on disease transmission risk. Still, they tended to hold neutral-to-positive attitudes, with most recognising bats as being important to ecosystem integrity and, as such, deserving of protection. Yet, urbanites reported diverging attitudes towards bats inhabiting their neighbourhoods (positive) or their yards/households (negative; [25]), once again illustrating the contextuality of attitudes.

Early education within families and in schools also plays a major role in modifying bat-related behaviours, and even attitudes. For example, Slovakian primary school children from urban contexts who knew more about bats held more positive attitudes towards them, while false beliefs arising from myth and folklore were associated with negative attitudes [27]. However, a general dislike of bats is not a strictly urban phenomenon. For instance, when Swiss students were shown images of wild animals, including the greater mouse-eared bat, *Myotis myotis*, they strongly preferred other species over bats, regardless of whether their schools were rural or urban [28].

4 Encounters Between Bats and Urbanites: A Change of Attitude?

A more general biophilia (referring to the human urge to affiliate with nature, *sensu* [29]) and associated positive feelings towards bats may be driving shifts in attitudes towards and beliefs about bats. Urbanites' low familiarity with nature in general, and bats in particular, makes bats potentially appealing in terms of curiosity and charisma, especially if human-bat interactions are mediated by experts (or physical barriers) and are thus perceived as safer. Signs of a shift in attitudes may include the spread of activities such as bat-related tourism in cities, including zoo exhibits and roost-emergence viewing, bat rescue efforts, and the installation of artificial roosts (e.g. bat boxes) as activities that bring people closer to bats.

Zoos are important constituents of traditional "green spaces" in many large cities and, as such, may enhance urbanites' experiences with – and awareness of – wildlife. Bats had long been mostly neglected as zoo animals, except for common vampire bats (*Desmodus rotundus*) and some flying foxes (*Pteropus* spp.), kept in several collections around the world as objects of curiosity, or ghost bats (*Macroderma gigas*), kept in several Australian zoos. Recently, zoos have added more bats to their collections and thus increased public attention to these mammals. As early as 1996, Bat Conservation International reported that 1.6 million people visited the "Lied Jungle" bat exhibit at the Henry Doorly Zoo, in central Omaha, Nebraska, USA, within one year of its opening. Similarly, in Europe, the Bat Cave at Hellabrunn Zoo in Munich, hosting Seba's short-tailed bats (*Carollia perspicillata*), opened in 1992. The advent of third-generation walk-through exhibits including bats has also improved the ability of zoos to raise awareness of wildlife (and bat) conservation and promote more positive perceptions of bats [30]. Zoos may also be excellent experimental settings to assess urbanites' feelings about bats, provided that husbandry practices ensure the welfare of individual bats in exhibits. For instance, children visiting a bat exhibit at the Brookfield Zoo, in Chicago, Illinois, USA, showed conflicting emotions towards fruit bats, i.e. fear and concern for their wellbeing [31].

The phenomenon of bats emerging from large colonies at dusk is also increasingly attracting public attention to urban bats. For example, among 17 large roosts used by Mexican free-tailed bats (*Tadarida brasiliensis*) in the Southwestern USA, two urban ones in Texas draw more than 65% of the 242,000 annual visitors to these roosts [32].

Another frequent type of human-bat interaction in cities occurs when people find grounded bats, usually injured adults or dependent juveniles that have fallen from their roosts [33]. In such cases, individual attitudes towards bats are the main predictor of the outcome, i.e. whether people rescue, ignore, or kill the bat. Despite generally negative attitudes towards bats being reported in the few studies that have assessed urbanites, positive behaviours such as rescuing bats in trouble seem relatively common in some cities. Yet, in other urban contexts, negative behaviours may also be strongly rooted as common habits. For example, the most frequent behaviours exhibited by people finding bats in their homes in Brazil were hunting and removal [34]. As such, wildlife rescue is a quantitatively relevant aspect of the bat-human interface (zone of interaction) in cities, suggesting that urbanites may engage in positive interactions such as rescue regardless of their general attitude. For example, each year, the Wildlife Rescue Centre of Rome, Italy, admits more than 300 bats per year, 95% of which come from Rome's city centre, leading to 3,548 human-bat encounters in the last 10 years (pers. obs.), and such rates are comparable with those in other locales [35, 36]. Such high rates of encounters suggest the important role of wildlife rescue initiatives in bat conservation, particularly in urban areas.

Wildlife rescue activities involve two main types of human-bat interactions: (1) members of the public finding a bat in distress and (2) staff and volunteers advising people and caring for bats. These interactions differ in the underlying perceptions and motivations of the people involved and in their potential contributions to conservation. In the first case, urbanites are motivated to rescue a bat (or have it rescued by professionals or volunteers), which presumably indicates a certain degree of general biophilia [30], and are usually keen to be advised on how to properly rescue the bat. Yet, people finding a bat in need of help generally lack knowledge of bats' basic biology and needs. For example, they frequently report having provided potentially harmful foods (e.g. honey, fruit, sugar to insectivores) to a bat before bringing it to a rescue centre, possibly a result of having viewed viral videos of flying foxes in Australian rescue centres. More rarely, they wrongly identify a bat as something else (e.g. bird, mouse or even frog, pers. obs.), a possible sign of disconnection from nature [37].

Rescue centres play an important role in addressing the poor knowledge and opinions that urbanites often have of bats. By bringing a bat in need of help to a centre, people may boost the chances that the bat survives [33] all while being offered proper support and information about bats that may further raise their appreciation for bats and promote their conservation. For example, the Valpredina Wildlife Rescue Centre, which is run by the WWF and admits >130 bats annually, represents an exemplary Italian case of management of the human-bat interface. The centre hosts a front office ("Sportello pipistrelli" – Life+ Project IPE 018 Gestire 2020) which, besides providing 24/7 support to people who find bats in trouble, promotes public involvement in conserving urban bat colonies. For example, staff members encourage those who rescue bat pups to search for the roost on their property and become "colony guardians". When urbanites enrol in the program, they are mentored by bat experts, who regularly provide information about bats in general and how to monitor "their" colonies, i.e. promoting a citizen-based monitoring approach and conscious cohabitation with bats. Empowering urbanites to protect bat colonies fosters an emotional bond between people and bats all while representing a durable way to protect urban roosts.

As for the motivations and attitudes of bat carers and volunteers, they remain poorly known. The only study to date [35] profiled carers of Australian flying foxes from a socio-demographic point of view, but only marginally explored their knowledge base and did not strictly focus on urbanites. This study reported that most bat carers were women between the ages of 30 and 50 years old, but detected no particular pattern with regard to previous knowledge and motivation to volunteer for bats. Moreover, volunteers appeared to be motivated by general altruism rather than by the drive to conserve bats.

Urbanites may also seek out contact with bats on their properties, in addition to casual encounters or volunteering. In recent years, bat boxes have gained popularity in urban green spaces. Urbanites set up bat boxes because they anticipate that doing so will suppress mosquitoes near their homes and/or to provide habitat for bats, partly thanks to large campaigns that have improved bats' "image" and increased urbanites' connection to bats (pers. obs.). For instance, the Natural History Museum of Florence launched a successful campaign ("Un Pipistrello per Amico", i.e. "Be a Bat's Friend) in 2006 [38]. Its goal was to disseminate accurate information about the ecological value of bats and foster direct public engagement in bat-conservation activities. The project involved active cooperation with one of Italy's main super-market chains, which sold more than 25,000 inexpensive bat boxes along with a range of bat-related gadgets. Citizens, schools, institutions, and associations installed bat boxes and monitored them using a standard data collection form. Bats colonised 40% of bat boxes, but the operation's biggest achievement was a general improvement in bat-related attitudes across Italy, e.g. with bat boxes becoming objects that were in high demand and commonly sold. Bat box campaigns, like other citizen science and conservation activities, offer an outstanding but still under-utilised potential to improve bats' public image and counter negative portrayals fostered by myths and misinformation. This example also highlights how important it is for experienced staff to provide supervision throughout projects involving the design and use of bat boxes, as these may easily become ecological traps if not properly designed or positioned [39].

Case Study: What Singaporeans Know and Feel About Bats

Here, we provide an example of an investigation into the attitudes and their antecedents towards bats by urbanites in a densely populated tropical city. This study, motivated by an interest in bats and belief that conservation must ultimately change human behaviours (see also [3]), sought to understand pre-dictors of behaviours towards bats in the understudied context of a large tropi-cal city, specifically focusing on knowledge and attitudes as antecedents (see also [40]).

The context was Singapore, a densely populated city-state whose popula-tion (5.64 million in 2018) is fully urbanised. By the turn of the millennium, many of Singapore's original bat species had become locally extinct [41], with most of the remaining species being locally threatened. Even though Singapore's bats need protection to avoid further declines and extinctions, removing bats (except for flying foxes, which are casual vagrants) is not ille-gal if one can prove they are damaging property. Based on this policy and lack of legal protection for resident species, one might infer that bats have no place in the city [42].

This mixed method study started with two focus group discussions (five participants each), whose outcomes were used to craft a quantitative survey of 106 Singapore residents, whose demographics were representative of the city's general population. Finally, individuals who expressed fear of bats were interviewed.

When asked what they know about bats, focus group discussion partici-pants mainly said that bats eat fruit, are nocturnal, and hang upside down. However, nearly one third of survey respondents described their knowledge as "low", even though 79% had encountered live bats (albeit rarely in most

(continued)

cases). In response to the statement that bats benefit humans, 23.6% agreed – pollination (32%) and pest control (32%) were the most cited benefits. Fewer (20.8%) agreed that bats cause problems – 68% of them mentioned health or disease issues. Finally, the mean score on a knowledge quiz was 7.17/12 – the most challenging question was the "bats are blind" myth, which 29.2% of respondents believed.

Three components of attitude were assessed. The first one was affect, and most of the focus discussion group participants expressed neutral attitudes, no opinion, or unclear emotions. Roughly equal numbers of survey respondents said they like (33.3%), dislike (31.1%), and neither like nor dislike (26.4%) bats – 34.9% said they find bats scary, even in images or videos. Most (62.2%) said they would not willingly live near bats, and 34.9% cared about their survival. As for the second, cognitive, component, while 66% of respondents agreed that bats are part of the environment, most disagreed with statements that they are attractive (54.7%) or charismatic (57.8%). Still, fewer than half agreed with any negative characterisations of bats (as unpredictable, disgusting, dangerous, or nuisances). Finally, social attitudes appeared to reflect societal expectations. In focus group discussions, participants varied in their level of support for protecting bats, but few were opposed to doing so. Most survey respondents agreed that bats have the right to exist (76.4%), are an important part of nature (74.5%), and should be conserved (68.9%), though only 39.6% said governments should fund this.

Religious or ethnic beliefs about bats were rare (8.5% of survey respondents). Three participants expressed the view of bats as auspicious – most Singaporeans are ethnically Chinese, and the Chinese character for bat (fú 蝠) is a homonym for luck (fú 福). Three viewed bats as God's creations, two said all creatures have intrinsic worth, and one said eating bats is *haram* (forbidden under Islam).

As to intrinsic predictors of cognitive attitudes, two predicted more positive ones. One was pet ownership, as in other studies (e.g. [7]), though why is unclear. Do people keep pets because they are animal lovers, or does owning pets promote love of animals? The other was positive prior experience (Wald's $X^2 = 10.1$) – again, the link is unclear. Animal lovers might be more observant and so report more bat interactions or may be more likely to describe interactions as positive.

The expectation that gender is predictive (based on findings that negative attitudes towards bats may be more common among females, e.g. [43]) was not supported by the data, leading to the hypothesis that sex-biased differences are minimised in a fully urbanised population that sees bats only rarely. Nor did knowledge of bats predict positive attitudes as it had in other studies (e.g. [43]). Perhaps low knowledge, as in Singapore, leads to weaker, less consistent, rather than negative, attitudes (see also [44]).

(continued)

In Singapore's densely built-up city, where bat sightings are rare, most residents may be effectively "blind to bats" [45]. This may be problematic for conservation. Upon encountering a familiar object, we retrieve the relevant affect and beliefs from memory to express an attitude towards it, but for a new object, we may rely on emotions [46], thus forming attitudes that resist cognitive persuasion [47]. Take, for instance, one interviewee whose fear of bats started after a bat flew near them. Teaching this person that bats are non-aggressive and are key to ecosystems could be futile or even backfire, due to cognitive dissonance [48], and with the negative affect formed first, they would probably disregard new information [47]. Using affect might change their mind, though one hurdle remains: we avoid thinking about things we fear [49]. Conversely, once a positive affect is formed, cognitive persuasion can reinforce it. Therefore, we suggest that outreach efforts not only present bats in a way that targets emotions but also (given the prevailing attitude that bats are not attractive) use aesthetically appealing portrayals rather than live bats (see also [3]).

Naturally, outreach can usefully present cognitive information on how bats benefit humans [4], but this approach may not always be contextually appropriate. For example, in Singapore's highly urbanised society, with no major agricultural sector, ecosystem services by bats do not yield economic benefits that most locals appreciate. Singapore hosts a diverse bat assemblage, which may potentially provide several services such as suppression of insect pests and pollination/seed dispersal of plants consumed by locals, yet such services are not yet quantified and unlikely to have high economic relevance. Indeed, when focus group discussion participants asked why bats should be protected and were told that bats pollinate durians (which are prized fruits in Singapore), this piqued their interest. Yet one countered, "Singapore does not produce durians", implying the fact's potential irrelevance locally. Still, if most respondents support conserving bats for other reasons and agree that bats are part of nature, perhaps outreach may be more effective by enhancing nature appreciation generally, rather than focusing on bats per se.

5   New Challenges

As evidenced by the reported examples, the human dimensions of bats in cities is a research topic with much to be learned. In most cases, knowledge of human behaviours towards bats (and their drivers) is anecdotal or based on single case studies at best. The contribution of social sciences to bat conservation is now widely recognised [50]. What is lacking is the identification and promotion of best practices and approaches to foster positive behaviours by urbanites towards bats, which, again, are needed to conserve them [4]. Negative perceptions of bats by urbanites may be

Fig. 10.2 Rescued bats, such as this European free-tailed bat (*Tadarida teniotis*) from Italy, offer great opportunities to take catchy pictures that showcase the cute or even charismatic aspects of bats, exploiting aesthetics in shaping people's attitudes towards bats. (Photo by Anna Maria Gibellini)

generally ameliorated by outreach activities that raise their knowledge of bats and by the efforts of Bat Conservation International (especially those with the capacity to directly involve people). Initiatives such as the International Year of the Bat and the International Bat Night (https://eurobats.org/international_bat_night) spread a globally unified message and allow local bodies and organisations (universities, agencies, associations, conservationists) to inform and involve urbanites under an international and authoritative framework. Such efforts may also represent a positive feedback mechanism to bat monitoring and conservation, as also suggested by the recent rise of volunteer-based monitoring schemes [51]. Experimental evidence strongly suggests an important role of aesthetics and direct experience, in addition to scientific information, in shaping people's attitudes towards bats [24], increasing the educational role of zoo collections and rescue centres (Fig. 10.2) as fundamental opportunities for urbanites to directly experience bats. Moreover, outreach actions, such as the applications of informative, aesthetically appealing panels at urban bat roosts, may facilitate acceptance of bats in buildings by urbanites, decreasing the risk of conflict and/or consequent damage to urban bats, as already evidenced in other contexts and taxa [52].

The role of bat conservationists has become even more important recently, given the need for outreach to mitigate the increase in negative attitudes and persecution stemming from media portrayals of bats in relation to the COVID-19 pandemic [50, 53]. With many bats, including species that occur in cities, being threatened, it is vital to provide efficient tools and consistent strategies to promote peaceful coexistence between bats and humans and public support for bat conservation.

Bat-related human dimension research is young and emerging, lacking systematic efforts to unify approaches and depict global trends until recently (e.g. [50], but see [54]). This situation presents many research opportunities to both social scientists and conservationists. One such opportunity is to address the relative lack of studies focusing specifically on the urban context – a worthwhile endeavour in the urban century, when more than half of all people already live in cities, which will house over two thirds of humanity by 2050. Fostering positive bat-related attitudes and norms among urbanites, e.g. by addressing the underlying cognitive and emotional drivers, may usefully help reduce persecution and other behaviours that may be detrimental to bat conservation.

We encourage human dimensions researchers delving into this area to engage in rigorous studies grounded in social sciences. Most fundamentally, studies should involve social scientists meaningfully and from the outset – thereby increasing the likelihood that appropriate conceptual frameworks and methodologies are applied and best social science practices are followed (see also [18]).

Literature Cited

1. Blair RB (2001) Birds and butterflies along urban gradients in two ecoregions of the United States: is urbanization creating a homogeneous fauna? In: Lockwood JL, McKinney ML (eds) Biotic homogenization. Springer, Boston, pp 33–56
2. Hwang YH, Jain A (2021) Landscape design approaches to enhance human–wildlife interactions in a compact tropical city. J Urban Ecol 7(1):juab007
3. Mascia MB et al (2003) Conservation and the social sciences. Conserv Biol 17(3):649–650
4. Kingston T (2016) Cute, creepy, or crispy—how values, attitudes, and norms shape human behavior toward bats. In: Voigt CC, Kingston T (eds) Bats in the anthropocene: conservation of bats in a changing world. Springer International Publishing, Cham, pp 571–595
5. Soga M, Gaston KJ (2016) Extinction of experience: the loss of human–nature interactions. Front Ecol Environ 14(2):94–101
6. Oh RRY et al (2020) No evidence of an extinction of experience or emotional disconnect from nature in urban Singapore. People Nat 2(4):1196–1209
7. Cox DTC et al (2017) Doses of neighborhood nature: the benefits for mental health of living with nature. Bioscience 67(2):147–155
8. Russo D, Ancillotto L (2015) Sensitivity of bats to urbanization: a review. Mamm Biol 80(3):205–212
9. Santini L et al (2019) One strategy does not fit all: determinants of urban adaptation in mammals. Ecol Lett 22(2):365–376
10. MacFarlane D, Rocha R (2020) Guidelines for communicating about bats to prevent persecution in the time of COVID-19. Biol Conserv 248:108650
11. Bennett NJ et al (2017) Conservation social science: understanding and integrating human dimensions to improve conservation. Biol Conserv 205:93–108
12. Fulton DC, Manfredo MJ, Lipscomb J (1996) Wildlife value orientations: a conceptual and measurement approach. Hum Dimens Wildl 1(2):24–47
13. Vaske JJ, Donnelly MP (1999) A value-attitude-behavior model predicting wildland preservation voting intentions. Soc Nat Resour 12(6):523–537
14. Whittaker D, Vaske JJ, Manfredo MJ (2006) Specificity and the cognitive hierarchy: value orientations and the acceptability of urban wildlife management actions. Soc Nat Resour 19(6):515–530

15. Ajzen I (2020) The theory of planned behavior: frequently asked questions. Hum Behav Emerg Technol 2(4):314–324
16. Jacobs MH, Vaske JJ, Roemer JM (2012) Toward a mental systems approach to human relationships with wildlife: the role of emotional dispositions. Hum Dimens Wildl 17(1):4–15
17. Castillo-Huitrón NM et al (2020) The importance of human emotions for wildlife conservation. Front Psychol 11(1277):1–11
18. Straka TM et al (2021) Human dimensions of bat conservation – 10 recommendations to improve and diversity studies of human-bat interactions. Biol Conserv. In revision
19. Kellert SR (1993) Values and perceptions of invertebrates. Conserv Biol 7(4):845–855
20. Knight AJ (2008) "Bats, snakes and spiders, Oh my!" how aesthetic and negativistic attitudes, and other concepts predict support for species protection. J Environ Psychol 28(1):94–103
21. López-Baucells A, Rocha R, Fernández-Llamazares Á (2018) When bats go viral: negative framings in virological research imperil bat conservation. Mammal Rev 48(1):62–66
22. Low M-R et al (2021) Bane or blessing? Reviewing cultural values of bats across the Asia-Pacific region. J Ethnobiol 41(1):18–34. 17
23. Kellert SR (1984) Urban American perceptions of animals and the natural environment. Urban Ecol 8(3):209–228
24. Boso À et al (2021) Understanding human attitudes towards bats and the role of information and aesthetics to boost a positive response as a conservation tool. Anim Conserv n/a(n/a)
25. Pérez B et al (2021) Design and psychometric properties of the BAtSS: a new tool to assess attitudes towards bats. Animals 11(2):244
26. Sexton NR, Stewart SC (2007) Understanding knowledge and perceptions of bats among residents of Fort Collins, Colorado. Fort Collins, p 22
27. Prokop P, Tunnicliffe SD (2008) "Disgusting" animals: primary school children's attitudes and myths of bats and spiders. Eurasia J Math Sci Technol Educ 4(2):87–97
28. Schlegel J, Rupf R (2010) Attitudes towards potential animal flagship species in nature conservation: a survey among students of different educational institutions. J Nat Conserv 18(4):278–290
29. Kellert SR, Wilson EO (1993) The biophilia hypothesis. Island Press, Washington, DC
30. Moss AG, Pavitt B (2019) Assessing the effect of zoo exhibit design on visitor engagement and attitudes towards conservation. J Zoo Aquarium Res 7(4):186–194
31. Kahn PH et al (2008) Moral and fearful affiliations with the animal world: children's conceptions of bats. Anthrozoös 21(4):375–386
32. Bagstad KJ, Wiederholt R (2013) Tourism values for Mexican free-tailed bat viewing. Hum Dimens Wildl 18(4):307–311
33. Serangeli MT et al (2012) The post-release fate of hand-reared orphaned bats: survival and habitat selection. Anim Welf 21(1):9–18
34. Santos NDJ et al (2019) Evaluation of bat-related knowledge, perceptions, and practices in an urban community: a strategy for conservation biology and health promotion. Braz J Biol Sci 6(13):347–358
35. Markus N, Blackshaw JK (1998) Motivations and characteristics of volunteer flying-fox rehabilitators in Australia. Anthrozoös 11(4):203–209
36. Kravchenko K et al (2017) Year-round monitoring of bat records in an urban area: Kharkiv (NE Ukraine), 2013, as a case study. Turkish J Zool 41:530–548
37. Zhang W, Goodale E, Chen J (2014) How contact with nature affects children's biophilia, biophobia and conservation attitude in China. Biol Conserv 177:109–116
38. Agnelli P et al (2011) Artificial roosts for bats: education and research. The "Be a bat's friend" project of the Natural History Museum of the University of Florence. Hystrix Ital J Mammal 22(1)
39. Bideguren G et al (2019) Bat boxes and climate change: testing the risk of over-heating in the Mediterranean region. Biodivers Conserv 28(1):21–35
40. Gifford R, Nilsson A (2014) Personal and social factors that influence pro-environmental concern and behaviour: a review. Int J Psychol 49(3):141–157

41. Lane DJW, Kingston T, Lee BPYH (2006) Dramatic decline in bat species richness in Singapore, with implications for Southeast Asia. Biol Conserv 131(4):584–593
42. Yeo J-H, Neo H (2010) Monkey business: human–animal conflicts in urban Singapore. Soc Cult Geogr 11(7):681–699
43. Prokop P, Fančovičová J, Kubiatko M (2009) Vampires are still alive: Slovakian students' attitudes toward bats. Anthrozoös 22(1):19–30
44. Tarrant MA, Bright AD, Ken Cordell H (1997) Attitudes toward wildlife species protection: assessing moderating and mediating effects in the value-attitude relationship. Hum Dimens Wildl 2(2):1–20
45. Lunney D, Moon C (2011) Blind to bats: Traditional prejudices and today's bad press render bats invisible to public consciousness. In: Law B et al (eds) The biology and conservation of Australasian Bats. Royal Zoological Society of New South Wales, pp 44–63
46. van Giesen RI et al (2015) Affect and cognition in attitude formation toward familiar and unfamiliar attitude objects. PLoS One 10(10):e0141790
47. Edwards K (1990) The interplay of affect and cognition in attitude formation and change. J Pers Soc Psychol 59(2):202–216
48. Festinger L (1962) Cognitive dissonance. Sci Am 207(4):93–106
49. Rocklage MD, Fazio RH (2018) Attitude accessibility as a function of emotionality. Personal Soc Psychol Bull 44(4):508–520
50. Rocha R, López-Baucells A, Fernández-Llamazares Á (2021) Ethnobiology of bats: exploring human-bat inter-relationships in a rapidly changing world. J Ethnobiol 41(1):3–17. 15
51. Barlow KE et al (2015) Citizen science reveals trends in bat populations: the National bat Monitoring Programme in Great Britain. Biol Conserv 182:14–26
52. Miller ZD et al (2018) Targeting your audience: wildlife value orientations and the relevance of messages about bear safety. Hum Dimens Wildl 23(3):213–226
53. Lu M et al (2021) Does public fear that bats spread COVID-19 jeopardize bat conservation? Biol Conserv 254:108952
54. Pérez B, Álvarez B, Boso A, Lisón F (2021) Design and psychometric properties of the BAtSS: a new tool to assess attitudes towards bats. Animals 11(2):244. https://doi.org/10.3390/ani11020244

Chapter 11
Urban Bats, Public Health, and Human-Wildlife Conflict

Christina M. Davy, Arinjay Banerjee, Carmi Korine, Cylita Guy, and Samira Mubareka

Abstract Coexistence of humans and bats in cities requires mitigation of two key sources of human-bat conflict: risk of zoonotic disease transmission and human concerns about cleanliness. Bats can transmit infectious diseases to humans, and mitigating this risk is an important challenge for both public health and bat conservation. Bat colonies in buildings (or adjacent to buildings) are often categorised as "nuisance wildlife" even when disease risk is low. These colonies can be noisy and create guano deposits that can be substantial and unsightly. Colonies of fruit bats may also feed on fruit grown for human consumption. In this chapter, we review perceived public health concerns around human-bat cohabitation and the factors that can increase or reduce the risk of disease transmission from urban bats to humans. We briefly review the importance of human dimensions in assessing the risk of zoonotic spillover and other bat-human conflict. We use two case studies (Boxes 11.1 and 11.2) to illustrate the implications of urban bats for human-wildlife conflict and public health: one on guano deposition by Egyptian fruit bats (*Rousettus aegyptiacus*) and the other on the risk of rabies exposure for humans cohabiting

C. M. Davy (✉)
Department of Biology, Carleton University, Ottawa, ON, Canada
e-mail: Christina.Davy@carleton.ca

A. Banerjee
Vaccine and Infectious Disease Organization, Department of Veterinary Microbiology, University of Saskatchewan, Saskatoon, SK, Canada

Department of Biology, University of Waterloo, Waterloo, ON, Canada

C. Korine
Mitrani Department of Desert Ecology, Ben-Gurion University of the Negev, Midreshet Ben-Gurion, Israel

C. Guy
Toronto, ON, Canada

S. Mubareka
Sunnybrook Research Institute and Department of Laboratory Medicine and Pathobiology, University of Toronto, Toronto, ON, Canada

© The Author(s), under exclusive license to Springer Nature Switzerland AG 2022   153
L. Moretto et al. (eds.), *Urban Bats*, Fascinating Life Sciences,
https://doi.org/10.1007/978-3-031-13173-8_11

with big brown bats (*Eptesicus fuscus*). Finally, we briefly consider key priorities for studies of bat-borne disease transmission in cities.

Keywords Zoonoses · Spillover · Wildlife disease · Synanthropy · Bat-borne viruses

1 Urban Bats and Zoonotic Diseases

Spillover of known and novel zoonotic diseases poses a major threat to global public health and world economies [1]. The risk of spillover is highest in regions with high mammalian diversity (typically tropical forest biomes) where land use and land cover changes drive the loss of wildlife habitat [2, 3]. In these regions, diverse wildlife species are also traded in live animal markets, where many species of livestock and/or wildlife may be brought from different jurisdictions and are often co-housed under stressful and crowded conditions [4]. Finally, urbanisation, the most dramatic form of land cover change, has been associated with increased spillover [5, 6]. Urbanisation increases contact between humans and wildlife because synanthropic species (wildlife that benefit from living near humans) typically rely on anthropogenic habitats (including buildings). Thus, urban expansion and intensification are explicitly associated with increased spillover risk from synanthropes, which include many species of bat [5, 6].

The number and severity of recent spillover events involving bats (e.g. severe acute respiratory syndrome coronavirus (SARS-CoV), Middle East respiratory syndrome coronavirus (MERS-CoV), Marburg virus (MARV), and Hendra virus (HeV) [3, 7]) illustrate the public health relevance of bat-borne pathogens. Horseshoe bats (*Rhinolophus* spp.) are reservoir hosts of close relatives of SARS-CoV-2 (the causative agent of COVID-19): SARS-CoV-2-related viruses have been identified in samples collected from wild *R. pusillus*, *R. stheno*, *R. affinis*, and *R. malayanus* [8]. Despite a flurry of research activity, the precise evolutionary and intermediate hosts of SARS-CoV-2 and mechanisms of spillover to humans remain enigmatic at the time of writing [8, 9].

Bats' physiological and immune adaptations often allow them to tolerate chronic, persistent viral infections [3, 10], although pathogen diversity varies interspecifically and geographically. Natural and experimental infections of bats with viruses rarely cause clinical signs of disease, even when co-infected with multiple viruses that are pathogenic to humans or livestock [11, 12]. The bats' adaptations make them reservoir hosts for a multitude of viruses that may cause illness if they spill over to other mammals [13]. Pathogen diversity tends to be higher in bat species that live longer and form larger colonies and may be higher in species with broad geographic distributions in the Eastern Hemisphere [14]. However, rigorous comparisons are limited by geographically uneven pathogen surveillance to date, and bat-borne pathogens (and hence spillover risk) also occur in regions with low bat diversity [e.g. 15].

Disease caused by emerging bat-borne viruses, such as the coronaviruses that cause SARS, MERS, and COVID-19, is driven by dysregulation of the host immune response, including dampened antiviral and exaggerated pro-inflammatory responses [3]. Studies of selected bat species reveal adaptations to control both virus-mediated modulation of the bats' antiviral defensive responses and the inflammatory responses typically mounted following viral infections [3]. Indeed, understanding the adaptations that allow bats to tolerate viruses that cause severe disease in other mammals may lead to the discovery of new therapeutic targets in other mammals, including humans and livestock.

Despite their remarkable immune responses, urban bats can still transmit pathogens, and understanding the urban ecology of viral spillover can help predict and mitigate disease transmission to people, livestock, and other urban wildlife. Urban bats may face increased physiological stress or pathogen exposures compared to their non-urban counterparts, and both factors may increase spillover risk. For example, little red flying foxes (*Pteropus scapulatus*) that are nutritionally or reproductively stressed have higher levels of Hendra virus replication [16]. Similarly, little brown bats (*Myotis lucifugus*) co-infected with *Myotis lucifugus* coronavirus (MyL-CoV) and the fungus *Pseudogymnoascus destructans*, which causes white-nose syndrome in hibernating bats, shed greater coronavirus loads than singly infected bats [17]. It remains unclear how well such studies can be generalised among urban and non-urban contexts, and future research should directly test anthropogenic stressors that may affect viral replication in bats, facilitating increased virus shedding and thus spillover into other mammals.

2 Health Risks of Bat-Human Cohabitation: To Humans and to Bats

In urban settings, as elsewhere, several factors determine interspecific pathogen transmission. These factors fall broadly into two categories: (1) drivers of exposure to an infectious pathogen, which is an essential first step for infection, and (2) biological determinants of infection. Here, we use different scenarios to highlight how exposure and host-pathogen biology contribute to spillovers and pose risks to human and wildlife health, with an emphasis on conditions common in cities. These scenarios focus on viruses because bat-human transmission of bacterial and fungal pathogens remains poorly understood [18]. Nevertheless, the principles we discuss apply to all zoonoses.

Exposure is determined largely by the human-animal interface. This interface may be narrow and sustained, with intense exposure to one reservoir or intermediate host (e.g. certain agricultural settings where camel producers are exposed to MERS [19]). The interface may also be broader and more intermittent (e.g. incursions into forests due to socio-economic necessity or geopolitical conflict resulting in human exposures to Ebola virus). Some exposures are easy to identify; others are difficult to detect but can be assessed retrospectively. For example, occupational exposures to filoviruses increase the risk of seropositivity. The adjusted odds ratio for seropositivity can be as high as 3.4 for miners who share their workspace with bats in areas where Marburg is endemic, compared to community controls [20]. In this example, close, sustained, and frequent contact with reservoir species in a poorly ventilated space likely contributes to high rates of exposure, seropositivity, and outbreaks. Other forms of exposure are more complex and multifactorial, as demonstrated for Nipah and Hendra viruses. Reduced connectivity among urban bat populations may reduce viral transmission, apparently decreasing exposure risk, while also generating populations of susceptible bats that ultimately drive more intense outbreaks when a virus is reintroduced locally [21].

Habitat loss associated with urbanisation also brings high densities of susceptible humans into direct contact with wild animals that have nowhere else to go. Planting of mango orchards in recently deforested areas placed bats (island flying foxes, *Pteropus hypomelanus*), pigs, humans, and Nipah virus in close proximity, leading to spillover events [22]. Although this example is from a rural area of Malaysia, similar processes during urbanisation could increase exposure risk for urbanites.

Where exposures occur, spillover risk varies with biological determinants such as host susceptibility (whether the host can become infected by the pathogen) and host competence (whether the infected host can transmit the pathogen to another susceptible host). For example, feline coronaviruses (FCoV), which are ubiquitous pathogens of domestic cats (*Felis catus*), do not spill over into humans despite constant, widespread exposure. It appears there are biological barriers to spillover, probably because humans do not have homologous cell attachment and entry determinants to which both FCoV biotypes can attach [23]. Similar biological transmission bottlenecks seem to prevent human infection by bat-borne coronaviruses (alphacoronaviruses and betacoronaviruses) [24]. However, growing human impacts on bat habitat by urban and agricultural development expand the interface between humans, livestock, and bats [2], increasing the probability of exposure to a coronavirus strain that can attach to and enter human cells. Such exposures can lead to spillover, and if humans are competent hosts for the particular virus, cities provide ideal conditions for sustained person-to-person transmission.

In cities, probabilities of both exposure and spillover increase because of high human populations and the increased likelihood that people will encounter synanthropic species that can transmit pathogens to them. Integration of epidemiological and biological data into risk assessment tools can also be adapted to support urban surveillance of bat disease. Recent, relevant initiatives include genome-based zoonotic risk assessment [25] and the web tool SpillOver [26], which integrates available scientific evidence with expert opinion to predict risk. Host surveillance can also be prioritised based on validated models. For example, ecological trait-based models recently revealed 47 previously unrecognised bat hosts of betacoronaviruses [27].

When bats and humans share buildings, they also share potential health risks. Given the new, global distribution of SARS-CoV-2, COVID-19 hotspots also represent areas of greater potential risk of spillback (anthropozoonosis or reverse zoonosis) to synanthropic bats that roost in buildings and therefore share air with humans in enclosed spaces. The SARS-CoV-2 virus has spilled back from humans into other mammals, including farmed European mink (*Mustela vison*) and American mink (*Neovison vison*), domestic cats, and some livestock and zoo animals [3]. At the time of writing, a range of wild species are confirmed as competent hosts for ancestral SARS-CoV-2 [e.g. 28–32]. However, host competence has only been experimentally confirmed in Egyptian fruit bats [*Rousettus aegyptiacus*; 33] and was refuted in big brown bats [*Eptesicus fuscus* 34], both of which often share space with humans.

Although in vivo experimental infections with SARS-CoV-2 have not yet been performed on other bat species, host susceptibility can also be inferred from studies of the receptor angiotensin-converting enzyme 2 (ACE2), which interacts directly with the viral spike protein [26]. A study of 46 bat species representing the major phylogenetic clades in Chiroptera used virus-host receptor binding and infection assays to examine ACE2 orthologues [33]. Synanthropic bats were no more susceptible than non-synanthropes to SARS-CoV or SARS-CoV-2, and host susceptibility was not phylogenetically constrained, but varied among congeners. Analyses of ACE2 in mammals have also been used to predict host capacity to transmit SARS-CoV-2 back to humans, suggesting low risk for *E. fuscus* but high risk for several *Rhinolophus* spp. and some flying foxes (*Pteropus* spp.) [34]. Unfortunately, emerging SARS-CoV-2 variants of concern (VOCs) have demonstrated expanded host tropism. For example, ancestral SARS-CoV-2 did not infect mice efficiently, but the beta VOC does [35–37]. In addition, both delta- and omicron-like SARS-CoV-2 isolates have been identified in white-tailed deer [*Odocoileus virginianus*; 32].

Indeed, the ability of SARS-CoV-2 VOCs to infect bat species, including *E. fuscus* and *R. aegyptiacus*, remains unknown. Nearctic bats carry a range of alphacoronaviruses, but did not evolve alongside betacoronaviruses such as SARS-CoV-2, prompting speculation that these bats may be particularly vulnerable to *Betacoronavirus* infection [38]. This hypothesis was not supported by ACE2 analyses, which found no association between host susceptibility to SARS-CoV-2 and a species' distribution [33]. Taken together, these studies indicate that the risk of SARS-CoV-2 spillback from humans to synanthropic bats should be assessed on a species-specific basis and re-examined periodically to account for emerging VOCs.

Finally, the urban wildlife trade raises added risks of pathogen spillover. Many cities contain some form of wildlife market, where urbanites can purchase live animals and their parts for various uses including food, medicine, souvenirs, and the pet trade [4]. The ethics of wildlife trade are complex and outside the scope of this chapter. However, the trade involves bringing many wild animals together for sale, exposing diverse species to each other's pathogens, and housing them under stressful and crowded conditions that will likely suppress immune function [39]. The risk to human health is well-documented [40], but the associated risks of spillover to free-ranging urban bats and other urban wildlife deserve further study.

3   Nuisance Wildlife or Welcome Guest?

Some cultures consider bats to embody luck and good fortune and tolerate or even celebrate bat roosts in buildings [41]. Other cultures consider bats to be nuisance wildlife or associated with misfortune. The human dimension aspects of urban bats are beyond the scope of this chapter, but we revisit them briefly here because human attitudes and behaviours towards urban bats and other wild animals are key drivers of human-wildlife conflict and the risk of zoonotic spillover. Urbanites' reluctance to host or live alongside colonies of bats may derive from the unsightly nature of guano deposits in attics or on exterior walls, the stigma associated with living with perceived pests, or concern about disease transmission [41, 42]. Thus, urban bat colonies may be subject to intentional killing under the guise of wildlife control or pest extermination, even when disease transmission is unlikely [43]. Prior to 2000, intentional killing of "nuisance" bats roosting in buildings was a common source of bat mortality in North America and Europe [44], and this threat remains unresolved [42].

Bat-borne pathogens, such as bat rabies, Australian bat lyssavirus, filoviruses, and paramyxoviruses, pose real risks to human health [45]. As such, ensuring that urbanites appreciate the risks of direct contact with bats is key to building bat-friendly cities while protecting human health [e.g. 49]. The emergence of COVID-19 has complicated this messaging as media coverage of the origins of COVID-19 has created inaccurate perceptions of disease risk from bats, leading to increased persecution of bats [47]. For example, a survey of residents of Arkansas, United States, found that strong negative perceptions of local bat species were associated with the belief that bats transmit SARS-CoV-2, even though local species are not implicated in the origins of COVID-19. In contrast, respondents were less concerned about rabies, which persists in populations of many urban and non-urban Nearctic bats [48].

Outreach may improve public awareness of bat diversity and the need for bat conservation. For example, a study of bat pollinators of durian (*Durio* spp.) suggests an economic incentive to conserve bats by demonstrating the ecosystem service they render [49]. Such examples are more difficult to find in cities, where ecosystem services provided by bats are less clear. However, outreach can help reduce spillover risk by informing urbanites about the risks of direct contact with bats, such as rabies in building-roosting *E. fuscus* in cities in the United States [48] or filoviruses in tree-roosting straw-coloured fruit bats (*Eidolon helvum*) in Ghanaian cities [46]. Importantly, outreach must be informed by culturally relevant social science methodology that considers the drivers of bat persecution or protection [41, 42, 47].

Box 11.1: Case Study – Fighting over Fruit and Faeces; The Case of Rousettus aegyptiacus

Spillover risk, actual or perceived, is not the only driver of human-bat conflict in cities. Competition for resources can also lead to conflict, as in the case of *R. aegyptiacus*. This bat occurs across the tropical and subtropical areas of Africa and the Eastern Mediterranean [50]. It is abundant in non-urban and urban habitats, including large cities, such as Cairo, Egypt, and Cape Town, South Africa. This species' cosmopolitan distribution, ability to successfully reproduce and thrive in urban habitats, and habit of feeding on fleshy fruits, including commercial crops [51], may all lead to major conflicts with humans.

Rousettus aegyptiacus are considered agricultural pests [51], and in some cities, they are also considered urban "pests" because they feed on fruit in urban gardens, dirty the walls of buildings with their guano [52], and are considered potential sources of human disease. *Rousettus aegyptiacus* is not thought to be a competent host for Nipah virus or the five known Ebolaviruses [7, 53], but like other pteropodids, these bats are potential reservoirs for coronaviruses [33], and they can carry Marburg virus [12]. As of this writing, we are not aware of confirmed cases of pathogen transmission from *R. aegyptiacus* to humans despite widespread overlap between humans and this species of bat, particularly in cities. Nevertheless, the perceived risks of competition for fruit crops and pathogen transmission have spurred culling of *R. aegyptiacus* populations by shooting and cave fumigation, leading some countries to enact legislation to protect this species [54]. For example, Israel banned cave fumigation in 1985, and *R. aegyptiacus* is legally protected by the law in Israel, Cyprus, and Turkey and is explicitly protected by the Agreement on the Conservation of Bats in Europe (https://www.eurobats.org/about_eurobats/protected_bat_species).

In this case study, we discuss two non-harmful ways in which conflict between humans and *R. aegyptiacus* can be mitigated; (1) by reducing damage to urban fruit crops, and (2) by discouraging colonies from settling in locations where unsightly guano deposits create a nuisance for urbanites and a potential disease risk [54].

Tree-Covering and Pruning to Prevent Fruit Damage

Tree covering is probably the most effective and bat-friendly method to avoid fruit damage in commercial orchards and can also be applied to urban fruit trees. Covering urban trees with woven, unframed nets that have a mesh size <2 mm or even <5 mm [54, 55] can protect fruit while preventing tangling by birds, lizards, and bats. On small or easily accessible trees, fruit protection bags may be also used [55]. When a tree is too tall to cover, it can be pruned to enable direct protection with netting, or alternate, indirect methods can be used to discourage the bats. For example, *Rousettus aegyptiacus* tend to carry fruit from the fruiting tree to temporary feeding roosts 5–30 m away (pine

(continued)

Box 11.1 (continued)

trees and other non-fruiting species [56]). Pruning these feeding roosts may also discourage bats from visiting the nearby fruiting tree. Finally, new neighbourhoods or city gardens can be planned with trees that do not bear fleshy fruit planted near buildings, to avoid attracting frugivores [57], and fruiting trees planted in locations farther from buildings.

Bat Deterrent Devices to Prevent Fruit Damage or Guano Deposition

Rousettus aegyptiacus can also be repelled from consuming fruit near the fruiting tree or defecating on exterior walls and cars using acoustic, visual, or chemical deterrents. Audible and ultrasonic acoustic deterrents differ with respect to the frequency and amplitude of sound they emit, the beam of the speaker, and the cost, but they may repel pteropodids and other frugivores [52, 58]. Ultrasonic deterrents typically operate across short distances because the sounds they produce attenuate quickly, and their effectiveness is unclear [54]. Playback of sounds audible to humans (e.g. gunshots and the sound of distressed bats) has also been used to deter pteropodids [54], but playback of loud, distressing noises is inappropriate in urban settings. *Rousettus aegyptiacus* can also habituate to the sounds of acoustic deterrents, so the efficiency of these devices may also be short-lived.

Chemical deterrents can include smoke, which may temporarily repel bats from fruiting trees, and strong odours, which may be taxon-specific or are generally unpleasant [54]. These are clearly not well suited to urban contexts as they would also be unpleasant or unsafe for urbanites. Visual deterrents based on lighting [54] may also be difficult to implement in cities because background light levels in cities tend to already be higher than in non-urban areas and because lights that would deter fruit bats may also be unpleasant for humans. Urban *R. aegyptiacus* may also be less impacted by light-based deterrents as they are already habituated to artificial light at night.

Well-designed experiments are needed to better understand the potential effects of such deterrents on the foraging behaviour of the fruit bats [e.g. 59], as well as on humans and other urban animals. However, these may be better suited to non-urban contexts, as the available acoustic, chemical, and visual deterrents are not taxon-specific and are likely to inadvertently impact urban humans, domestic animals, and wildlife, as well as *R. aegyptiacus*. Thus, discouraging bats by netting and pruning fruiting trees (as described above) may be the most appropriate mitigation in urban areas.

Box 11.2: Case Study – Eptesicus fuscus as House Guests
Eptesicus fuscus are widely distributed in North America, where they are relatively common in cities and often form large colonies in buildings [60]. They are typically excellent house guests – they prefer not to use the same space as human inhabitants, and it is not uncommon for people to be unaware that they share a building with bats. Still, *E. fuscus* roosts pose potential disease risks to humans in at least two ways. First, large accumulations of guano can enable growth of *Histoplasma capsulatum* var. *capsulatum*, a soil-associated endemic and dimorphic fungus that can cause histoplasmosis when microconidia are inhaled [43]. Second, *E. fuscus* can carry rabies virus (RABV) in North America [61] and transmit rabies to humans through direct bat-human contact and through intermediate hosts (wild and domestic mesocarnivores). The risk of exposure to both diseases is minimised when urbanites understand how they are transmitted and how to avoid them.

Mitigating Risk of Histoplasmosis in Homes Shared with E. fuscus
Guano deposits in enclosed, dry spaces such as attics can provide suitable growth conditions for *H. capsulatum*. Microconidia are aerosolised when guano is disturbed (e.g. while removing deposits), which increases inhalation risk [43, 62]. This fungus is not a bat-guano specialist; it can also grow on other nitrogen sources, including droppings of domestic poultry [63]. Thus, histoplasmosis risk is not limited to cohabitation with (or exposure to) bats. Nevertheless, people living with any large colony of bats that create contained guano deposits should be informed about histoplasmosis so they can take appropriate precautions.

Fortunately, minimising exposure to *H. capsulatum* is relatively straightforward. Removing guano deposits from enclosed spaces regularly (annually) can prevent build-up of large amounts of guano. Building-roosting *E. fuscus* typically select enclosed interior spaces, such as attics, that humans rarely enter, which effectively isolates their guano deposits from the rest of the home. When entering these spaces, humans can wear respirators to prevent inhalation of *H. capsulatum* spores as the deposits are disturbed [64]. Ideally, guano removal is (1) performed by professionals who have effective personal protective equipment and (2) completed in the winter, after resident bats have moved elsewhere to hibernate. This can minimise direct human-bat contact and thus reduce the risk of transmission of other diseases, including rabies.

Transmission of Rabies by Eptesicus fuscus: Low Probability, High Stakes
The rabies virus (RABV) belongs to the family Rhabdoviridae and genus *Lyssavirus* and causes fatal encephalitis in infected mammals. Rabies causes tens of thousands of human deaths globally each year, most of which occur in Asia and Africa and result from bites by infected dogs [45]. The situation in North America is very different – cases in humans are rare but are overwhelmingly associated with bat-borne strains of RABV [15, 65].

(continued)

Box 11.2 (continued)

Eptesicus fuscus is one of the most commonly submitted species for rabies testing in North America because, thanks to its building-roosting habit, it is encountered more often than most other species of bat. The incidence of RABV in *E. fuscus* appears low: a mere 3.6% of 62,997 *E. fuscus* tested from 2010 to 2015 in the United States were carrying the virus [65]. However, this sample is not representative (i.e. individual *E. fuscus* that are tested for rabies have already behaved atypically enough to be sent for testing), so true population prevalence is likely much lower. These data can be further parsed by RABV strain as the various strains of bat-borne RABV are not evenly represented in human cases.

Spillover of rabies from Nearctic bats is not clearly associated with synanthropic species of bats. The most commonly implicated strains in human cases of bat rabies in North America are associated with silver-haired bats (*Lasionycteris noctivagans*), tricolored bats (*Perimyotis subflavus*), and Brazilian free-tailed bats (*Tadarida brasiliensis*). The first two species are rarely synanthropic and infrequently encountered by humans. However, strains from these species could be transmitted by synanthropes such as *E. fuscus*. Of 232 RABV sequences isolated from *E. fuscus* in Canada, 5 were assigned to strains associated with other bat species [61]. Isolates from *E. fuscus* submitted for rabies testing in the United States provide further evidence for occasional spillover: 94.8% of 116 genotyped samples belonged to strains associated with *E. fuscus*, but the rest originated in other species [65]. Limited genotyping of RABV strains from rabies-positive bats limits conclusions about the rate of RABV spillover among bat species [65], but RABV strains from other species can clearly spill over to *E. fuscus*, which can transmit these strains directly to humans, and to domestic or wild mammals that can also transmit to humans. The apparently low rate of spillover is intriguing, because strains associated with *E. fuscus* are rarely transmitted to humans despite their abundant urban colonies [65, 66].

Minimising the risk of RABV transmission between *E. fuscus* and humans is critical, because human rabies is almost always fatal [15]. One mitigation strategy is public education. First, people must be instructed never to handle bats with bare hands. When bats enter the parts of a building used by humans, leaving doors and windows open usually allows the bats to leave of their own accord. If a bat is acting strangely or simply does not leave, it may be safely picked up while wearing thick leather work gloves, or gently scooped up using a dustpan and container, and released outside. Finally, the public should be informed that bites from *E. fuscus* (and many other species) can be practically invisible even if the teeth have broken the skin. If a bite occurs, or a bat enters a room where someone is sleeping or incapacitated, and a bite may have occurred unnoticed, medical attention should be sought immediately [15].

4 Reducing Risk of Zoonotic Spillovers from Urban Bats

Cities bring together high densities of humans, domestic animals, and wildlife animals [5, 21], bringing urban bats into contact with a myriad of pathogens they may not encounter in non-urban areas. Conversely, humans sharing buildings or green spaces with urban bats may be exposed to bat-borne pathogens, whether directly or indirectly through contact with bats, bat guano, or domestic animals that may be intermediate hosts [22]. Clarifying these transmission pathways can identify exposure hotspots and inform mitigation to reduce spillover risk [2].

Effective, humane deterrents for bat species that are perceived as nuisances or pests could reduce human-bat contact, but public education may provide a more effective solution by helping people to appreciate living with bats [42]. Of course, to safely live with bats (or any urban wildlife), people must know about potential public health risks and how to mitigate them. An essential message is that eradicating or culling bats is unlikely to reduce, and can even exacerbate, risks of disease transmission [42]. Conservation of bats in cities, with a focus on outreach and risk mitigation (rather than eradication), is therefore a public health imperative.

Future research should continue to untangle disease dynamics among urban bats, recognising that bat-borne pathogens can pose a risk to human health, while human-to-bat spillover (e.g. potential spillback of SARS-CoV-2) or human-mediated transmission of other pathogens to bats (e.g. dispersal of *P. destructans*) can pose a threat to bat health. There are challenges in advancing this work, and we highlight two in particular. First, procedural variation among surveillance studies can limit the power of systematic reviews or meta-analyses to detect temporal or spatial trends. Standardising detection methods, assay interpretation, and diagnostic criteria for diseases among studies will increase the long-term impact of surveillance. For example, studies exploring whether bats are a principal reservoir for Ebola virus are highly heterogeneous in sample type, assay (molecular or serological), and taxonomic and demographic representation (species, sex, age), and are frequently based on small sample sizes with limited statistical power. Establishing best practices would facilitate inter-study comparison and enable more powerful meta-analyses. An example of how this could be done on a global scale comes from surveillance for avian influenza in wild birds (http://www.fao.org/AVIANFLU/en/manuals.html), but this approach involves establishing reference centres instead of relying on sporadic and variable research protocols. Minimum metadata sets that include host traits and ecological context are essential, as demonstrated in a recent study that used ecological trait-based models to reveal previously unrecognised bat hosts for betacoronaviruses [27].

The importance of standardised surveillance intensity and effort is also illustrated by our case study of RABV prevalence in *E. fuscus*. Rabies is deadly in humans, and bat rabies is linked to most human cases in North America [15]. Yet, big knowledge gaps include prevalence of the various strains in *E. fuscus* (current sampling is not representative) and the factors driving spillover among bat species. Standardised, long-term surveillance of existing pathogen diversity can characterise temporal and regional variation in prevalence of zoonoses and their causative agents and identify the drivers of outbreaks and spillovers.

Finally, although surveillance studies provide an essential baseline, research should also explicitly address the gap between identifying existing pathogen diversity and understanding the drivers of bat-human spillover in the urban context [2, 67]. This requires sustained investment in coordinated, long-term, collaborative approaches that include community organisations, academic and government scientists, public health agencies, and social scientists, as well as wildlife rehabilitators, zoos, non-governmental organisations (NGOs), and community science initiatives. Such a comprehensive, One Health framework can not only benefit public and animal health but also raise awareness of the interconnectedness of bats, humans, and the biosphere, and of the benefits of coexistence.

Acknowledgements We thank two anonymous reviewers and the editor of this chapter for their helpful comments and suggestions.

Literature Cited

1. World Health Organization (2005) The control of neglected zoonotic diseases: a route to poverty alleviation: report of a joint WHO/DFID-AHP meeting. WHO, Geneva
2. Plowright RK et al (2021) Land use-induced spillover: a call to action to safeguard environmental, animal, and human health. Lancet Planet Heal 5:e237–e245
3. Irving AT et al (2021) Lessons from the host defences of bats, a unique viral reservoir. Nature 589:363–370
4. Huong NQ et al (2020) Coronavirus testing indicates transmission risk increases along wildlife supply chains for human consumption in Viet Nam, 2013–2014. PLoS One 15:2013–2014
5. McFarlane R et al (2012) Synanthropy of wild mammals as a determinant of emerging infectious diseases in the Asian-Australasian region. EcoHealth 9:24–35
6. Eskew EA, Olival KJ (2018) De-urbanization and zoonotic disease risk. EcoHealth 15:707–712
7. Jones M et al (2015) Experimental inoculation of Egyptian Rousette bats (*Rousettus aegyptiacus*) with viruses of the Ebolavirus and Marburgvirus genera. Viruses 7:3420–3442
8. Zhou H et al (2021) Identification of novel bat coronaviruses sheds light on the evolutionary origins of SARS-CoV-2 and related viruses. Cell 184:4380–4391.e14
9. Banerjee A (2021) Unraveling the zoonotic origin and transmission of SARS-CoV-2. Trends Ecol Evol 36:180–184
10. Brook CE, Dobson AP (2015) Bats as "special" reservoirs for emerging zoonotic pathogens. Trends Microbiol 23:172–180
11. Munster VJ et al (2016) Replication and shedding of MERS-CoV in Jamaican fruit bats (*Artibeus jamaicensis*). Sci Rep 6:1–10
12. Guito JC et al (2021) Asymptomatic infection of Marburg virus reservoir bats is explained by a strategy of immunoprotective disease tolerance. Curr Biol 31:257–270.e5
13. Letko M et al (2020) Bat-borne virus diversity, spillover and emergence. Nat Rev Microbiol 18:461–471
14. Guy C et al (2020) The influence of bat ecology on viral diversity and reservoir status. Ecol Evol 10:5748–5758
15. Fenton MB et al (2020) Bat bites and rabies: the Canadian scene. Facets 5:367–380
16. Plowright RK et al (2008) Reproduction and nutritional stress are risk factors for Hendra virus infection in little red flying foxes (*Pteropus scapulatus*). Proc R Soc B Biol Sci 275:861–869

17. Davy CM et al (2018) White-nose syndrome is associated with increased replication of naturally persisting coronaviruses in bats. Sci Rep 8:1–12
18. Allocati N et al (2016) Bat–man disease transmission: zoonotic pathogens from wildlife reservoirs to human populations. Cell Death Discovery 2:1–8
19. Alshukairi AN et al (2018) High prevalence of MERS-CoV infection in camel workers in Saudi Arabia. MBio 9:1–10
20. Nyakarahuka L et al (2020) A retrospective cohort investigation of seroprevalence of Marburg virus and ebolaviruses in two different ecological zones in Uganda. BMC Infect Dis 20:1–9
21. Plowright RK et al (2011) Urban habituation, ecological connectivity and epidemic dampening: the emergence of hendra virus from flying foxes (*Pteropus spp.*). Proc R Soc B Biol Sci 278:3703–3712
22. Kessler MK et al (2018) Changing resource landscapes and spillover of henipaviruses. Ann N Y Acad Sci. https://doi.org/10.1111/nyas.13910
23. Jaimes JA, Whittaker GR (2018) Feline coronavirus: insights into viral pathogenesis based on the spike protein structure and function. Virology 517:108–121
24. Banerjee A et al (2019) Bats and coronaviruses. Viruses 11:7–9
25. Mollentze N et al (2021) Identifying and prioritizing potential human infecting viruses from their genome sequences. PLoS Biol 19:1–25
26. Fischhoff IR et al (2021) Predicting the zoonotic capacity of mammals to transmit SARS-CoV-2. Proc R Soc B Biol Sci 288:20211651
27. Becker DJ et al (2022) Optimising predictive models to prioritise viral discovery in zoonotic reservoirs. Lancet Microbe. https://doi.org/10.1016/s2666-5247(21)00245-7
28. Oude Munnink BB et al (2021) Transmission of SARS-CoV-2 on mink farms between humans and mink and back to humans. Science 371:172–177
29. Griffin BD et al (2021) SARS-CoV-2 infection and transmission in the North American deer mouse. Nat Commun 12:1–10
30. Hale VL et al (2021) SARS-CoV-2 infection in free-ranging white-tailed deer. Nature 602:481–486
31. Schlottau K et al (2020) SARS-CoV-2 in fruit bats, ferrets, pigs, and chickens: an experimental transmission study. Lancet Microbe 1:e218–e225
32. Hall JS et al (2021) Experimental challenge of a North American bat species, big brown bat (*Eptesicus fuscus*), with SARS-CoV-2. Transbound Emerg Dis. https://doi.org/10.1111/tbed.13949
33. Yan H et al (2021) ACE2 receptor usage reveals variation in susceptibility to SARS-CoV and SARS-CoV-2 infection among bat species. Nat Ecol Evol 5:600–608
34. Grange ZL et al (2021) Ranking the risk of animal-to-human spillover for newly discovered viruses. Proc Natl Acad Sci U S A 118:20210413
35. Radvak P et al (2021) SARS-CoV-2 B.1.1.7 (alpha) and B.1.351 (beta) variants induce pathogenic patterns in K18-hACE2 transgenic mice distinct from early strains. Nat Commun 12:1–15
36. Pan T et al (2021) Infection of wild-type mice by SARS-CoV-2 B.1.351 variant indicates a possible novel cross-species transmission route. Signal Transduct Target Ther 6:420
37. Shuai H et al (2021) Emerging SARS-CoV-2 variants expand species tropism to rodents. EBioMedicine 73:103643
38. Olival KJ et al (2020) Possibility for reverse zoonotic transmission of SARS-CoV-2 to free-ranging wildlife: a case study of bats. PLoS Pathog 16:1–19
39. Smith KM et al (2017) Wildlife hosts for OIE-listed diseases: considerations regarding global wildlife trade and host–pathogen relationships. Vet Med Sci 3:71–81
40. Eskew EA, Carlson CJ (2020) Overselling wildlife trade bans will not bolster conservation or pandemic preparedness. Lancet Planet Heal 4:e215–e216
41. Kingston T (2016) Cute, creepy, or crispy—how values, attitudes, and norms shape human behavior toward bats. In: Bats in the anthropocene: conservation of bats in a changing world. Springer, Cham, pp 571–595
42. Frick WF et al (2020) A review of the major threats and challenges to global bat conservation. Ann N Y Acad Sci 1469:5–25

43. Voigt CC et al (2016) Bats and buildings: the conservation of synanthropic bats. In: Bats in the Anthropocene: conservation of bats in a changing world. Springer, Cham, pp 427–462
44. O'Shea TJ et al (2016) Multiple mortality events in bats: a global review. Mammal Rev 46:175–190
45. Cleaveland S, Hampson K (2017) Rabies elimination research: juxtaposing optimism, pragmatism and realism. Proc R Soc B Biol Sci 284:20171220
46. Gbogbo F, Kyei MO (2017) Knowledge, perceptions and attitude of a community living around a colony of straw-coloured fruit bats (*Eidolon helvum*) in Ghana after Ebola virus disease outbreak in West Africa. Zoonoses Public Health 64:628–635
47. Rocha R et al (2021) Bat conservation and zoonotic disease risk: a research agenda to prevent misguided persecution in the aftermath of COVID-19. Anim Conserv 24:303–307
48. Sasse DB, Gramza AR (2021) Influence of the COVID-19 pandemic on public attitudes toward bats in Arkansas and implications for bat management. Hum Dimens Wildl 26:90–93
49. Sheherazade et al (2019) Contributions of bats to the local economy through durian pollination in Sulawesi, Indonesia. Biotropica 51:913–922
50. Kultzer E (1979) Ecology and geographical range in the fruit-eating cave bat genus *Rousettus* Gray 1821 – a review. Bonner Zool Beiträge 30:233–275
51. Korine C et al (1999) Is the Egyptian fruit bat *Rousettus aegyptiacus* a pest in Israel? An analysis of the bat's diet and implications for its conservation. Biol Conserv 88:301–306
52. Mickleburgh SP et al (1992) Old World fruit bats. An action plan for their conservation. IUCN, Gland
53. Seifert SN et al (2020) *Rousettus aegyptiacus* bats do not support productive Nipah virus replication. J Infect Dis 221:S407–S413
54. Aziz SA et al (2016) The conflict between pteropodid bats and fruit growers: species, legislation and mitigation. In: Kingston T, Voigt C (eds) Bats in the Anthropocene: conservation of bats in a changing world. Springer, Cham, pp 377–426
55. Tollington S et al (2019) Quantifying the damage caused by fruit bats to backyard lychee trees in Mauritius and evaluating the benefits of protective netting. PLoS One 14:1–13
56. Izhaki I et al (1995) The effect of bat (*Rousettus aegyptiacus*) dispersal on seed germination in eastern Mediterranean habitats. Oecologia 101:335–342
57. Peters VE et al (2016) Using plant–animal interactions to inform tree selection in tree-based agroecosystems for enhanced biodiversity. Bioscience 66:1046–1056
58. Richards GC (2002) The development of strategies for management of the flying-fox colony at the Royal Botanic Gardens, Sydney. In: Managing the Grey-headed flying-fox. Royal Zoological Society of New South Wales, Mosman, pp 196–201
59. Harten L et al (2020) The ontogeny of a mammalian cognitive map in the real world. Science 369:194–197
60. Agosta SJ (2002) Habitat use, diet and roost selection by the big brown bat (*Eptesicus fuscus*) in North America: a case for conserving an abundant species. Mammal Rev 32:179–198
61. Nadin-Davis SA et al (2010) Spatial and temporal dynamics of rabies virus variants in big brown bat populations across Canada: footprints of an emerging zoonosis. Mol Ecol 19:2120–2136
62. Bartlett PC et al (1982) Bats in the belfry: an outbreak of histoplasmosis. Am J Public Health 72:1369–1372
63. Bilgi C (1980) Pulmonary histoplasmosis: a review of 50 cases. Can Fam Physician 26:225–22530
64. Morris T, Coleman L (2017) Acceptable management practices for bat control activities in structures in Georgia – a guide for nuisance wildlife control operators. White-nose Syndrome Conservation and Recovery Working Group, U.S. Fish and Wildlife Service, Hadley, MA
65. Pieracci EG et al (2020) Evaluation of species identification and rabies virus characterization among bat rabies cases in the United States. J Am Vet Med Assoc 256:77–84
66. Walker FM et al (2021) Relatedness and genetic structure of big brown bat (*Eptesicus fuscus*) maternity colonies in an urban-wildland interface with periodic rabies virus outbreaks. J Wildl Dis 57:303–312
67. Combs MA et al (2021) Socio-ecological drivers of multiple zoonotic hazards in highly urbanized cities. Glob Change Biol 28:1705–1724

Chapter 12
Ecosystem Services by Bats in Urban Areas

Danilo Russo, Joanna L. Coleman, Leonardo Ancillotto, and Carmi Korine

Abstract Preserving biodiversity and the ecosystem services (ES) it provides is vital to sustainability. With over half of all people living in cities, urban ES play an especially important role. Bats are the most speciose mammalian group in many cities and may provide a variety of important ES. This chapter explores the available literature and provides unpublished information on bats' urban ES, covering insectivory, pollination, seed dispersal, and bat-related tourism.

We found that although research interest in bat-mediated ES has grown considerably, ES by urban bats have been relatively neglected. Twenty-two studies on various continents that used molecular identification of prey remains in bat droppings revealed a substantial consumption of urban pests, including 'nuisance' insects, such as drain flies and mosquitos, and species that bite or induce allergic reactions. Bats also consume the same species of mosquitos that are vectors of the West Nile virus and malaria, phlebotomine sandflies that transmit leishmaniasis, as well as insects that damage stored products. Ecosystem services rendered by phytophagous bats (pollination and seed dispersal) in urban areas are poorly known but potentially important. However, urbanisation might hinder the movement of bats and thus limit

Supplementary Information The online version contains supplementary material available at [https://doi.org/10.1007/978-3-031-13173-8_12].

D. Russo (✉) · L. Ancillotto
Laboratory of Animal Ecology and Evolution (AnEcoEvo), Dipartimento di Agraria,
Università degli Studi di Napoli Federico II, Portici, Naples, Italy
e-mail: danrusso@unina.it

J. L. Coleman
Department of Biology, Queens College at the City University of New York,
Flushing, NY, USA

C. Korine
Mitrani Department of Desert Ecology, Ben-Gurion University of the Negev,
Midreshet, Ben-Gurion, Israel

the provision of such ES. The few studies available fail to prove that phytophagous bats in urban areas mediate plant recruitment, yet there is some anecdotal evidence that they do. Urban bats also provide cultural ES, such as bat-related tourism, which in some cases generate considerable revenue. We highlight a significant gap in research on urban ES by bats that must be filled given its expected importance. We call for studies to document and quantify such ES, especially studies that adopt rigorous methods, such as DNA metabarcoding in faecal analysis, exclosure experiments to study insect suppression, or bat-mediated pollination, and studies that provide evidence of the role of bats in increasing fruit set and/or plant recruitment. Raising awareness of bat-mediated urban ES might help convince the public that bats are essential components of urban biodiversity and are worth conserving.

Keywords Biodiversity · Chiroptera · Cities · Ecosystem services · Insectivory · Pollination · Seed dispersal

1 Introduction

Ecosystem services (ES), here defined as the benefits provided to people by ecosystems and their components, bear a value that is well-recognised at national and international scales. Preserving biodiversity, i.e., the source of ES, and its provisions is therefore regarded as a top priority in sustainability policy [1]. Although only ca. 3% of global land is classified as urban, more than half of all people live in cities, a fraction that is projected to rise to two-thirds by 2050 [2]. This trend underscores the increasing importance of urban landscapes to humankind. Although conversion of natural and semi-natural habitats to urban land uses is seen as a major cause of loss of ES [2], there is growing research interest in ES provided by urban biodiversity [3].

   Although many mammal species are threatened by urbanisation, some tolerate, or even thrive in, urban habitats [4]. Within urban mammal communities, bats are an important group, with several species that persist not only in remnants of natural habitats but also in the most built-up areas, where they may roost in buildings or forage near streetlamps [5]. These include many bat species, which, like other wildlife, may provide both ES and ecosystem disservices [ED; 6]. Whereas ED, such as conflicts caused by the presence of bats in buildings and sometimes zoonoses (diseases that originate in non-human animals), are well-known [7], there is less knowledge of the role of bats as providers of urban ES.

   In this chapter, we provide a state-of-the-art picture of ES delivered by urban bats by analysing the available literature and complementing it with unpublished information on bat pollination in Southeast Asia. The specific ES we address are insectivory, pollination, and seed dispersal by bats and the economic and cultural benefits of bat-related tourism.

   To assess the representation of bats in the literature on urban ecosystem services (UES), we performed a Web of Science (WoS) search, with 'urban' AND 'ecosystem service' as topics (Appendix 1). Of 830 relevant records (identified by scanning all titles and abstracts), 99.5% were published since 2005, i.e., the same year as the Millennium Ecosystem Assessment, which popularised the term 'ecosystem service'. Interest has risen exponentially since then and is dominated by China (which accounts for more than one-third of all authors and study sites), followed by the United States (24% of all authors, 14% of study sites). Only 31 studies were done in sub-Saharan Africa, which is urbanising faster than any other region [8]. Most research aims to map the distribution of/change in ES, analyse policy related to UES, valuate UES, or assess human perceptions of UES. Where direct, animal-mediated ES are concerned, nearly all studies focus on insect pollinators, with only one study [9] on bats.

A similar WoS search but with 'bats' AND 'ecosystem service' as topics (Appendix 2) identified 114 relevant records. Again, research interest has grown exponentially since 2005. Studies of specific bat-mediated ES mainly focus on bats as pest control agents – as may be expected given that most bats are insectivorous [10]. Seed dispersal is the next most assessed ES, followed by pollination. In this case, the geographical bias pertains to authorship only, with 38% of authors based in the United States, followed by the United Kingdom and Germany, each with more than 15% of authors. Most study localities (71%) have been in areas of high biodiversity and many bat species, i.e. the tropics and subtropics, but only ten studies included urban sites.

Of course, this literature search does not reveal the full scope of work on ES by urban bats – records are limited to those with 'ecosystem service' in the topics. As described below, searching for 'urban' AND 'pollination' AND 'bats', for example, yields a different set of records, including some from cities. Still, it is fair to say that urban bats have been largely overlooked as providers of ES. This oversight may seem surprising considering that among mammal orders, Chiroptera is the most speciose one in many cities [4, 5]. Furthermore, by virtue of their unmatched dietary diversity and long-distance foraging behaviours, their importance as providers of ES is substantial if undervalued [11]. To help fill this gap, we explore ES provided by bats in urban areas, presenting the latest knowledge and highlighting research directions.

1.1 Insectivory

Insectivorous bats in natural and semi-natural habitats supply key intermediate (regulating) ES, such as the suppression of pest arthropods (Fig. 12.1a). The importance of these services by bats across agricultural landscapes is well-understood [12] and has been economically valuated [13]. In contrast, very little information is available about the importance of these services in urban areas. Nevertheless, what information does exist suggests that the benefits may be considerable.

For instance, the role of bats as predators of mosquitos may be locally important [14] (Figs. 12.2 and 12.3) and substantially reduce egg production [15], suggesting that bats exert top-down regulation on mosquito populations, and do not merely

Fig. 12.1 Urban-relevant ecosystem services rendered by bats: (**a**) A greater mouse-eared bat (*Myotis myotis*) – a species known to occur in urban areas – feeding on a rose chafer (*Cetonia aurata*), a horticultural pest (photo by Merlin Tuttle). (**b**) A dog-faced fruit bat (*Cynopterus brachyotis*) flying off with the fruit of a yellow stem fig (*Ficus fistulosa*) in Singapore, potentially acting as a seed disperser (photo by Chan Kwok Wai). (**c**) The emergence of Mexican free-tailed bats (*Tadarida brasiliensis*) from the Congress Avenue Bridge in Austin, Texas, USA, draws onlookers each night and is the city's most popular attraction (photo by Merlin Tuttle). (**d**) A cave nectar bat (*Eonycteris spelaea*) foraging on the nectar of a banana (*Musa* sp.) flower in Singapore and thus potentially acting as a pollinator (photo by Nick Baker)

suppress them. Bats might thus help mitigate the transmission of mosquito-borne diseases in cities [15], including some that cause significant morbidity and mortality among urbanites worldwide.

Bats also consume other perceived pests. These include chironomids [14; present analysis (Figs. 12.2 and 12.3)], which cause economic damage, nuisance, and health issues in cities [16], and moths, e.g., the oak processionary (*Thaumetopoea processionea*) [17], whose caterpillars, besides damaging trees, release urticating hairs that irritate people and domesticated animals. Finally, Gould's wattled bats (*Chalinolobus gouldii*) in Perth, Australia, were recently found to consume hundreds of insect species, but especially lepidopterans, 40% of which are deemed pests [9]. Otherwise, studies of ES rendered by insectivorous bats in cities are practically non-existent. We present the first analysis of urban pests from studies on the diets of bats from various continents.

Previous work adopting morphological analysis of food remains in bat droppings provides limited evidence of pest consumption because insect fragments large enough to offer sufficient taxonomic resolution are rare, especially for moths, which comprise many agricultural pests [12]. However, modern molecular techniques are highlighting that urban bats deliver this important ES. For instance, using DNA

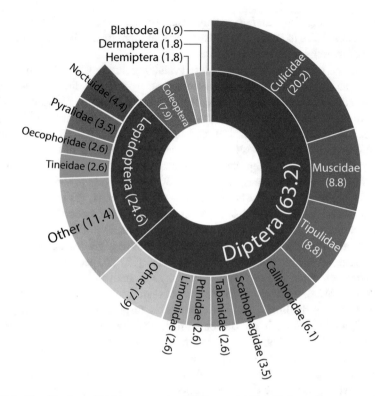

Fig. 12.2 Classification of 114 insect species (1008 records) identified as urban pests and documented in the diets of bats (*n* = 22 publications) into orders and families. Numbers in parentheses are percentages of species in each taxon

Fig. 12.3 Relationships with humans of insect species (n = 114) occurring in the diet of the bats taken from 22 publications and identified as urban pests. Only occurrence frequencies >2.5% are shown. The numbers in parentheses are percentages

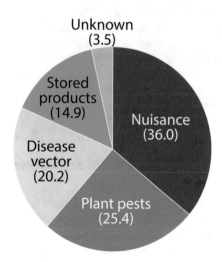

metabarcoding, Aguiar et al. [18] recently analysed faecal samples from five house-dwelling bat species in Brazil and determined that more than half of the 83 molecular taxonomic units were from pest insects.

We adopted a conservative approach to evaluating the possible ES of urban bats, by collecting data on diets of insectivorous bats from 22 studies that used DNA metabarcoding to identify insects at least to the level of genus. These studies concerned Australia, Africa, Europe, Asia, and the Neotropics and included 1008 records of insects. We used a handbook [19] and online databases (CABI, 2019: Invasive Species Compendium: https://www.cabi.org/isc/ and Plant Pest of the Middle East: (http://www.agri.huji.ac.il/mepests/) to delineate five groups of insect taxa based on their relationships with humans, as proposed by the literature: (1) agricultural pests, which also include known pests of plants in urban habitats, (2) storage pests, (3) potential disease vectors, (4) nuisance pests, and (5) neutral association (non-pests). We considered categories 2–4 as primarily urban pests, even though some agricultural pests also cause problems in urban green spaces (e.g. gardens, parks).

We could not unambiguously categorise 8.1% (82) of species due to a lack of information on their ecology and biology. Our analysis revealed that 46.2% ($n = 465$), 32.5% ($n = 328$), and 13.1% ($n = 114$) of insect species were categorised as natural pests, agricultural pests, and urban pests, respectively. Nearly half (45.6%) of the categorised insects in our database are pests in agricultural or urban habitats. Documented taxa of urban pests consumed by bats belonged to 44 families, in 6 orders (especially Diptera and Lepidoptera), with the most frequently consumed families being Culicidae, Muscidae, and Tipulidae (Fig. 12.2).

Urban arthropods in the database included species whose pest status we classified into four broad categories (Fig. 12.3). More than one-third could be considered species whose bites are painful or provoke allergic reactions and 'nuisance' insects, such as drain flies (Psychodidae) and mosquitos (Culicidae), which can become pests in anthropogenic aquatic habitats such as sewage treatment facilities. One-quarter of the species are documented pests of various plants (trees, shrubs, crops) that are cultivated in urban green spaces and their surroundings. One-fifth of species are disease vectors, including species of mosquitos that transmit West Nile virus and malaria and phlebotomine sandflies (Psychodidae) that transmit leishmaniasis. The rest (15%), mainly moths and beetles, are species known to damage stored products including foodstuffs, household materials, or woollen clothing.

1.2 Pollination and Seed Dispersal

Phytophagous bats in the Old World (Pteropodidae) and New World (Phyllostomidae) tropics and subtropics are key pollinators and seed dispersers for many plant species (Fig. 12.1b, d). In several cases, bats are the only animals known to fulfil these ecological roles. Approximately 500 plant species rely solely on bats for pollination, exemplified by tight mutualisms, such as those between *Heliconia solomonensis* and Woodford's fruit bat (*Melonycteris woodfordi*) in the Solomon Islands or

Centropogon nigricans and the tube-lipped nectar bat (*Anoura fistulata*) in Ecuador [see also 20]. Although plants' relationships with seed-dispersing bats are rarely exclusive (obligate), the seeds of hundreds of plant species are dispersed by bats [21]. Thus, bats are vital mediators of outcrossing and succession, perhaps especially where, thanks to their long-distance foraging behaviour, they disperse plants in degraded landscapes that may have lost other animal dispersers [see also 22 and others therein].

However, the above ecological roles of bats only equate to ES if there is a quantifiable (ideally, valuated) benefit to humans, and convincing demonstrations that phytophagous bats provide ES are scarce generally, let alone in cities [see also 22]. A Web of Science search (on 20 December 2020; Appendix 3), with 'urban' AND 'pollination' AND 'bats' in the topics, yielded eight unique, relevant records (after discarding those that were doubled or on other topics). We found one more record opportunistically from citations within these records or our knowledge of the literature. A similar search on the same date (substituting 'seed' for 'pollination') yielded 19 records. Of these papers, one on pollination and five on seed dispersal generated no useful knowledge on hypothesised or realised ES and/or were not about urban bats. Thus, it is evident that links between urbanisation and ES by phytophagous bats are still poorly understood and represent a research frontier.

Research on the role of urban bats as pollinators and seed dispersers is worth pursuing. If bats are important agents of dispersal for plants in cities, then they are indirectly implicated in the ES rendered by urban plants, including providing food, fibre, and wildlife habitat, carbon sequestration, mitigating urban-related abiotic changes (e.g., urban heat islands, pollution, altered hydrology), and improving human-nature relationships [see also 3]. However, urbanisation could also undermine these ES via extreme habitat fragmentation and the prevalence of artificial light and noise pollution, all of which could hinder movement by bats [see also 23]. For example, for urban calabash trees (*Crescentia cujete*), which are only pollinated by Pallas's long-tongued bat (*Glossophaga soricine*), outcrossing declines with distance between individuals on a campus in Recife, Brazil – a finding interpreted as being indicative of reduced habitat connectivity for bats [23]. However, pinpointing exactly which urban attributes limit connectivity (e.g., impervious cover, light, noise) should be done through spatial, landscape-scale analyses and/or experimental manipulation.

Yet, adequate spatial (or temporal) replication and experimentation are rare among studies of ES by urban phytophagous bats. Rather, most studies have documented diet and/or foraging behaviour in one or a few sites, using morphological identification of plant parts collected from fur and/or faeces or direct observation. Ecosystem services provided by bats have then been inferred based on which plants bats eat or visit. Such studies often find that urban bats have broad diets and/or readily eat exotic species [e.g. 24–28, and others therein] – tendencies also revealed by spatially replicated, urban-gradient studies in Southeast Asia [29, 30] and a systematic review of studies on diets of great fruit-eating bats (*Artibeus lituratus*) in Brazil [31]. By promoting increased dietary generalism and making exotic plants more available to bats, urbanisation may produce a loss of plant species that exhibit bat pollination syndrome, as shown by studies of floral resources (e.g. in Thailand [29] and Brazil [32]), and raise competition among native plant species for effective agents of dispersal [33]. These native plants may have high conservation value and/or deliver important ecosystem services.

Experimental work demonstrates that the consumption of seeds by urban fruit bats does not hinder or even improves germination outcomes [e.g. 34, 35]. For example, in Singapore, the lesser dog-faced fruit bat (*Cynopterus brachyotis*) forages on 33 species, including heritage trees and pioneer plants (Fig. 12.1b), and seeds remain viable after gut passage [30]. Moreover, this bat species moves seeds well away from parent plants. These findings suggest a role in dispersing plants of conservation and/or cultural importance.

However, the above-mentioned studies do not prove that urban fruit bats mediate plant recruitment because they do not document the location of the seed rain and/or the fate of seeds. One study [36] found more tree seedlings beneath sacred figs (*Ficus benghalensis*) in settlements than in open or riparian areas and showed that these figs attract visits by Indian flying foxes (*Pteropus giganteus*). But they attract birds too. As such, it remains unclear which animal species are effectively dispersing seeds. Likewise, studies of flower-visiting urban bats have not documented whether these visits lead to successful pollination. Therefore, evidence that urban bats are effective pollinators or seed dispersers needs strengthening.

Yet, there is some anecdotal evidence of bat pollination constituting an effective ES. In Singapore (as throughout Southeast Asia), durian (*Durio* spp.) is considered the 'King of Fruits', though it is all imported, mainly from Malaysia [37]. The fruit is so prized that each year, mainstream media publish articles on current and forecasted prices to help locals find the best deals. But instead of buying durian, some locals collect fallen fruits in parks or on other public land, even though doing so is illegal. Bats are the main pollinators of wild and semi-wild durian, and the most effective species, the cave nectar bat, *Eonycteris spelaea* [e.g. 38, 39], is locally common – the large roost mentioned by Leong and Chan [40] contained >1,400 individuals in 2015 and is <1 km from one of Singapore's few durian groves (unpubl. data). In such groves, durians develop without hand pollination – which points to the role of animal pollinators, almost certainly *E. spelaea* – and are eaten by locals.

Urban phytophagous bats may also provide ecosystem disservices, although the distinction between 'services' and 'disservices' is subjective. For example, in Singapore, *C. brachyotis* eats and moves seeds of the Jamaican cherry (*Muntingia calabura*) [30], an exotic species deemed invasive, but whose fruits are eaten by several native birds (pers. obs.). Thus, a birder who enjoys watching these birds might disagree with a conservationist as to whether this ecological role is welcome.

1.3   Bat Tourism

Urban bats may provide cultural ecosystem services, which are defined as the nonmaterial benefits (e.g., recreation, spirituality, etc.) that humans derive from biodiversity. For instance, Mexican free-tailed bats (*Tadarida brasiliensis*) roost under the Congress Avenue Bridge in Austin, Texas, USA (Fig. 12.1c), and the opportunity to watch their nightly emergence draws more than 242,000 visitors per year, generating a conservatively estimated revenue of 6.5 million USD [41]. Bat-related tourism may also have positive knock-on repercussions for the conservation of bats and biodiversity generally. Bat tourism may raise public awareness of the importance of bats and enhance support for conservation, especially among urbanites, whose dissociation from nature may adversely affect environmental governance and sustainable urbanisation [3]. For example, visiting a bat cave sanctuary in the Philippines was found to increase visitors' willingness to conserve [42]. And each year, a bat centre in the town of Barrea, Abruzzo, Italy, receives thousands of tourists who come to watch the televised activity of a large maternity colony of greater horseshoe bats (*Rhinolophus ferrumequinum*) in a building [43]. In summer 2020 alone, an estimated 12,000 people visited the bat centre (A. Scarnecchia, pers. comm.). This case demonstrates how it is possible to protect a major bat colony while also providing economic benefits to locals, education, and enjoyable experiences to tourists as well as behavioural research [43]. Bat tourism may also involve bat exhibits, festivals, and interactive interfaces that allow the public to engage with and learn about urban bats (e.g., bat walks, listening to echolocation calls, public bat surveys, etc.) – such activities may successfully engage urbanites and improve their knowledge and perspectives of bats [44].

2   Conclusions

Cities are now home to most of humanity, i.e., the beneficiaries of ES. This is especially true in the world's most rapidly urbanising nations, which are also the lowest-income ones [8] and where urbanites may therefore be most reliant on ES [45]. Therefore, understanding the ES rendered by urban bats can have major economic and sustainability implications. Our comprehensive review of the literature highlights a lack of research attention to the ES (and ED) rendered by urban bats. This is striking because cities that have lost large fractions of their pre-settlement bat faunas may still retain several species that have diverse functional traits and are often abundant – key attributes that underpin ES [46].

Evidence from non-urban habitats speaks to the potentially large value of these urban ES, which is expected to grow in the urban century. For instance, as rural-to-urban migration and the environmental crisis compromise conventional agriculture, urban farming becomes more important and stands to improve not only food security but also nutrition and income for urbanites [47]. The plants best suited to urban agriculture [e.g. *Brassica* spp.; 47] are not pollinated by bats but do provide a foraging habitat for certain bats, e.g. Kuhl's pipistrelle (*Pipistrellus kuhlii*), especially near human settlements [48].

Another key trend is the confluence of urbanisation and climate change exacerbating the risk of certain vector-borne diseases. The role of bats in mitigating this risk is exemplified by the case of dengue fever (the world's fastest-growing vector-borne disease) – whose causative agent is a virus transmitted by *Aedes* mosquitos (especially *A. aegypti*) that mainly occur in cities and whose range should expand with climate change [49]. Here, bats may provide a two-pronged ES. First, they could suppress the vectors – indeed, urban bats do consume mosquitos [9]. Second, bats, especially commensal species, appear to be dead-end hosts for the virus [50], thereby potentially reducing total infection pressure.

In this century, there is also a growing emphasis on biophilic cities to mitigate numerous urban-related abiotic changes, promote urban sustainability and resilience, and enhance urbanites' access to nature [3 and others therein]. Urban bats play key roles in these initiatives, as evidenced above. They pollinate and disperse seeds of cultivated and wild plant species, consume horticulturally important insect pests, and offer people the chance to observe wildlife.

However, most of the evidence for bat-mediated ES comes from non-urban systems, leaving a large research gap. Therefore, we call for rigorous studies in cities to document, quantify, and valuate these services, including the use of DNA metabarcoding techniques for faecal analysis that are exclusively designed for urban bats [as done by 18]. Ideally, studies focusing on pollination (and/or insect control) will involve experimentation (e.g., exclosure tests) to isolate the roles of bats from those of other animals and address potential confounders [see also 12]. Investigations of pollination and/or seed dispersal should ideally demonstrate that the involvement of bats ultimately increases fruit set and/or plant recruitment. Moreover, the cultural ES rendered by urban bats should be explored using established social science frameworks and techniques, despite ongoing challenges as far as defining categories of cultural ES and identifying the most appropriate methods to evaluate them [51].

As we approach global tipping points of biodiversity loss and climate change, the importance of garnering human support for sound conservation cannot be overstated. With most policy decisions made by urbanites (who also constitute most of humanity), there is an urgent need to increase public appreciation for bats (like all components of biodiversity) so that people will act swiftly and meaningfully to protect them. We argue not that raising awareness of bat-mediated urban ES is enough to accomplish this goal, but rather than doing so at least stands to persuade most people, whose motivations to conserve are grounded in utilitarian ethics.

Appendices

Appendix 1

Results of a Web of Science (WoS) search of the literature with 'urban' AND 'ecosystem service' as topics, performed on 5 May 2021.

The worksheet 'WoS savedrecs' displays raw results, with references sorted by year before assigning them unique serial numbers (column A).

We obtained country data (column D) for the worksheet 'Study locations' by scanning abstracts and (where the location was not evident from abstracts) papers to ascertain where the research was conducted.

The 'ES categories' worksheet classifies references into 16, author-defined, types of ecosystem services (row 1) that were investigated.

The 'WoS categories' and 'Research areas' worksheets contain the WoS graphical outputs depicting the 25 most common subject categories and research areas (defined by WoS), respectively.

The worksheet entitled 'years' contains a chart depicting growth in the literature over time, i.e. publications per year, with a best-fit exponential trend line and the associated R2 value – the data to construct this chart come from column N of the 'WoS savedrecs' worksheet.

The worksheets entitled 'Authors' and 'WoS countries (authors)' contain the graphical outputs from WoS, depicting the 25 most commonly represented authors and national affiliations of authors, respectively.

Appendix 2

Results of a Web of Science (WoS) search of the literature with 'bats' AND 'ecosystem service' as topics, performed on 5 May 2021.

The worksheet 'WoS ES + bats complete' displays the raw results, with references, ordered the same way that WoS returned records.

The worksheet entitled 'WoS ES + bats refined' only lists the 114 records that we established were relevant after scanning the abstracts and/or papers. We assigned each paper a dichotomised score indicating whether it investigated ecosystem services in an urban context (1) or not (0), in column D, and identified the locality (column E) where the research was done. Columns T through AA classify records into the six typical categories of bat-mediated ecosystem services that were investigated.

The 'Localities' worksheet is where we tabulated the locality data mentioned above to establish the geography of the research. The 'WoS categories' and 'Research areas' worksheets contain the WoS graphical outputs depicting the 25 most common subject categories and research areas (defined by WoS), respectively.

The worksheet entitled 'years' contains a chart depicting growth in the literature over time, i.e. publications per year, with a best-fit exponential trend line and the associated R2 value – the data to construct this chart come from column L of the 'WoS ES + bats refined' worksheet.

The worksheets entitled 'Authors' and 'WoS countries (authors)' contain the graphical outputs from WoS, depicting the 25 most represented authors and national affiliations of authors, respectively.

Appendix 3

Results of literature searches related to three types of ecosystem services rendered by bats.

The worksheets entitled 'Urban + pollination + bats' and 'Urban + seed + bats' list records returned by Web of Science (WoS) during topic searches performed on 20 December 2020, plus others identified opportunistically (from citations within these records or our knowledge of the literature). The records in red font in the 'Urban + seed + bats' worksheet are redundant, i.e. also returned by the WoS search of urban AND pollination AND bats in the topics.

The worksheet entitled 'Urban + insect + bats + control' lists records returned by a WoS search of these terms in topics performed on 5 May 2021.

Literature Cited

1. Griggs D et al (2013) Sustainable development goals for people and planet. Nature 495(7441):305–307
2. McDonald RI et al (2020) Research gaps in knowledge of the impact of urban growth on biodiversity. Nat Sustain 3(1):16–24
3. Tan PY et al (2020) A conceptual framework to untangle the concept of urban ecosystem services. Landsc Urban Plan 200:103837
4. Santini L et al (2019) One strategy does not fit all: determinants of urban adaptation in mammals. Ecol Lett 22(2):365–376
5. Jung K, Threlfall CG (2016) Urbanisation and its effects on bats – a global meta-analysis. In: Voigt CC, Kingston T (eds) Bats in the anthropocene: conservation of bats in a changing world. Springer, Cham, pp 13–33
6. Lyytimäki J et al (2008) Nature as a nuisance? Ecosystem services and disservices to urban lifestyle. Environ Sci 5(3):161–172
7. Voigt CC et al (2016) Bats and buildings: the conservation of synanthropic bats. In: Voigt CC, Kingston T (eds) Bats in the anthropocene: conservation of bats in a changing world. Springer, Cham, pp 427–462
8. United Nations, D.o.E.a.S.A., Population Division (2019) World urbanization prospects: The 2018 revision (ST/ESA/SER.A/420). UN, New York
9. Burgar JM, Hitchen Y, Prince J (2021) Effectiveness of bat boxes for bat conservation and insect suppression in a Western Australian urban riverine reserve. Austral Ecol 46(2):186–191

10. Jones G, Rydell J (2003) Attack and defense: interactions between echolocating bats and their insect prey. In: Kunz TH, Fenton MB (eds) Bat ecology. The University of Chicago Press, Chicago
11. Ghanem SJ, Voigt CC (2012) Increasing awareness of ecosystem services provided by bats. Adv Study Behav 44:279–302
12. Russo D, Bosso L, Ancillotto L (2018) Novel perspectives on bat insectivory highlight the value of this ecosystem service in farmland: research frontiers and management implications. Agric Ecosyst Environ 266:31–38
13. Boyles JG et al (2011) Economic importance of bats in agriculture. Science 332(6025):41–42
14. Puig-Montserrat X et al (2020) Bats actively prey on mosquitoes and other deleterious insects in rice paddies: potential impact on human health and agriculture. Pest Manag Sci 76(11):3759–3769
15. Reiskind MH, Wund MA (2009) Experimental assessment of the impacts of northern long-eared bats on ovipositing Culex (Diptera: Culicidae) mosquitoes. J Med Entomol 46(5):1037–1044
16. Broza M (2008) Chironomids as a nuisance and of medical importance. In: Capinera JL (ed) Encyclopedia of entomology. Springer, Dordrecht, pp 860–862
17. Garin I et al (2019) Bats from different foraging guilds prey upon the pine processionary moth. PeerJ 7:e7169
18. Aguiar LMS et al (2021) Going out for dinner—the consumption of agriculture pests by bats in urban areas. PLoS One 16(10):e0258066
19. Robinson WH (2005) Urban insects and arachnids: a handbook of urban entomology. Cambridge University Press, Cambridge
20. Fleming TH, Geiselman C, Kress WJ (2009) The evolution of bat pollination: a phylogenetic perspective. Ann Bot 104(6):1017–1043
21. Muscarella R, Fleming TH (2007) The role of frugivorous bats in tropical forest succession. Biol Rev 82(4):573–590
22. Regolin AL et al (2020) Seed dispersal by Neotropical bats in human-disturbed landscapes. Wildl Res 48(1):1–6
23. Diniz UM, Lima SA, Machado ICS (2019) Short-distance pollen dispersal by bats in an urban setting: monitoring the movement of a vertebrate pollinator through fluorescent dyes. Urban Ecosyst 22(2):281–291
24. Lim VC et al (2018) Pollination implications of the diverse diet of tropical nectar-feeding bats roosting in an urban cave. PeerJ 6:e4572
25. Corlett RT (2005) Interactions between birds, fruit bats and exotic plants in urban Hong Kong, South China. Urban Ecosystems 8(3):275–283
26. Rollinson DJ, Jones DN (2002) Variation in breeding parameters of the Australian magpie Gymnorhina tibicen in suburban and rural environments. Urban Ecosyst 6(4):257–269
27. Abedi-Lartey M et al (2016) Long-distance seed dispersal by straw-coloured fruit bats varies by season and landscape. Glob Ecol Conserv 7:12–24
28. Gulraiz TL et al (2016) Role of Indian flying fox Pteropus giganteus Brunnich, 1782 (Chiroptera: Pteropodidae) as a seed disperser in urban areas of Lahore, Pakistan. Turk J Zool 40(3):417–422
29. Sritongchuay T, Hughes AC, Bumrungsri S (2019) The role of bats in pollination networks is influenced by landscape structure. Glob Ecol Conserv 20:e1001
30. Chan AAQ et al (2020) Diet, ecological role and potential ecosystem services of the fruit bat, Cynopterus brachyotis, in a tropical city. Urban Ecosyst 24(2):251–263
31. Laurindo RD, Vizentin-Bugoni J (2020) Diversity of fruits in Artibeus lituratus diet in urban and natural habitats in Brazil: a review. J Trop Ecol 36(2):65–71
32. Oliveira MTP et al (2020) Urban green areas retain just a small fraction of tree reproductive diversity of the Atlantic forest. Urban For Urban Green 54:126779
33. Gelmi-Candusso TA, Hämäläinen AM (2019) Seeds and the city: the interdependence of zoochory and ecosystem dynamics in urban environments. Front Ecol Evol 7(41):20190301

34. de Figueiredo RA et al (2008) Reproductive ecology of the exotic tree Muntingia calabura l. (Muntingiaceae) in southeastern Brazil. Revista Arvore 32(6):993–999
35. Nakamoto A et al (2007) Feeding effects of Orii's flying-fox (Pteropus dasymallus inopinatus) on seed germination of subtropical trees on Okinawa-Jima Island. Tropics 17(1):43–50
36. Caughlin TT, Ganesh T, Lowman MD (2012) Sacred fig trees promote frugivore visitation and tree seedling abundance in South India. Curr Sci 102(6):918–922
37. Ketsa S (2018) Durian—*Durio zibethinus*. In: Rodrigues S, de Oliveira Silva E, de Brito ES (eds) Exotic fruits. Academic Press, London, pp 169–180
38. Aziz SA et al (2017) Pollination by the locally endangered island flying fox (Pteropus hypomelanus) enhances fruit production of the economically important durian (Durio zibethinus). Ecol Evol 7(21):8670–8684
39. Sheherazade, Ober HK, Tsang SM (2019) Contributions of bats to the local economy through durian pollination in Sulawesi, Indonesia. Biotropica 51(6):913–922
40. Leong TM, Chan KW (2011) Bats in Singapore – ecological roles and conservation needs. In: Proceedings of Nature Society, Singapore's conference on 'nature conservation for a sustainable Singapore'. Singapore
41. Bagstad KJ, Wiederholt R (2013) Tourism values for Mexican free-tailed bat viewing. Hum Dimens Wildl 18(4):307–311
42. Tanalgo KC, Hughes AC (2021) The potential of bat-watching tourism in raising public awareness towards bat conservation in the Philippines. Environ Challenges 4:100140
43. Ancillotto L, Venturi G, Russo D (2019) Presence of humans and domestic cats affects bat behaviour in an urban nursery of greater horseshoe bats (Rhinolophus ferrumequinum). Behav Process 164:4–9
44. Kaninsky M, Gallacher S, Rogers Y (2018) Confronting people's fears about bats: combining multi-modal and environmentally sensed data to promote curiosity and discovery. In: Proceedings of the 2018 designing interactive systems conference. Association for Computing Machinery, Hong Kong, pp 931–943
45. Suich H, Howe C, Mace G (2015) Ecosystem services and poverty alleviation: a review of the empirical links. Ecosyst Serv 12:137–147
46. Díaz S et al (2013) Functional traits, the phylogeny of function, and ecosystem service vulnerability. Ecol Evol 3(9):2958–2975
47. Clinton N et al (2018) A global geospatial ecosystem services estimate of urban agriculture. Earth's Future 6(1):40–60
48. Kahnonitch I, Lubin Y, Korine C (2018) Insectivorous bats in semi-arid agroecosystems – effects on foraging activity and implications for insect pest control. Agric Ecosyst Environ 261:80–92
49. Murray NEA, Quam MB, Wilder-Smith A (2013) Epidemiology of dengue: past, present and future prospects. Clin Epidemiol 5:299–309
50. Vicente-Santos A et al (2017) Neotropical bats that co-habit with humans function as dead-end hosts for dengue virus. PLoS Negl Trop Dis 11(5):e0005537
51. Cheng X et al (2019) Evaluation of cultural ecosystem services: a review of methods. Ecosyst Serv 37:100925

Chapter 13
The Big Picture and Future Directions for Urban Bat Conservation and Research

Krista J. Patriquin, Lauren Moretto, and M. Brock Fenton

Abstract Urbanisation modifies natural environments, creating light, noise, air, and water pollution, which may impact bat physiology, ecology, and behaviour. The vast variation in the physical and behavioural characteristics of bats makes it difficult to predict how each species will be affected by urbanisation. It appears that urban-dwelling bats commonly present general physiological and behavioural adaptations to urban environments: an "urban syndrome". Like other urban-dwelling mammals, bats may also move into and out of urban areas to feed, roost, and hibernate. However, the impacts of urban living on the survival and reproductive success of the few urban bats studied to date remain equivocal as responses are species-specific. In some instances, bat fitness appears to be higher in urban areas. In other instances, fitness is reduced in urban areas, suggesting they may be ecological traps for some bats. Additional species-specific research and tracking could improve our knowledge of urban-dwelling bats, which could better inform management actions that support bats. This could be supplemented by establishing clear definitions of urban environments and gradients of urbanisation across studies.

Keywords Urban syndrome · Urban mammals · Conservation action · Research directions · Urban gradient

K. J. Patriquin (✉)
Department of Biology, Saint Mary's University, Halifax, NS, Canada
e-mail: krista.patriquin@smu.ca

L. Moretto
Vaughan, ON, Canada

M. B. Fenton
Department of Biology, University of Western Ontario, London, ON, Canada

1 Urban Bats: Adaptation or Tolerance?

Urbanisation imposes several stressors on bats, such as light, noise, air, and water pollution, and changes to roosting and foraging resources [[1] (see Chaps. 1, 6, 7, 8, and 9)]. These stressors can affect habitat use, reproductive success, and survival. However, responses of bats to urbanisation cannot be easily generalised [2], as they may vary within and among species due to differences in energetic and thermal demands associated with age, sex, and reproductive condition (see Chap. 1). Nevertheless, urban bats are generally flexible in traits relating to roosting, diet, echolocation, and social structure (see Chaps. 1, 3, 5, 6, 7, and 9). Together, these traits may comprise an "urban syndrome" (see Chap. 1).

It is unclear if an urban syndrome is adaptive for bats. Adaptive traits promote an individual's survival and lifetime reproductive success, but there is limited evidence of fitness benefits gained by urban-dwelling bats. Bats may be at higher risk of mortality as they fly through cities to feed in patches of green space (see Chaps. 1, 3, 7, 8, and 9). Bats roosting in artificial structures may succumb to overheating (see Chap. 6), which can also affect bats roosting in the open during extreme heat events (Chap. 7). Urban bats may also carry greater parasite burdens than non-urban-dwelling bats, which may affect body condition, survival, and reproductive success, but these relationships vary with host species, demographics, season, and parasite species (see Chap. 4). With a few exceptions [e.g. Kuhl's pipistrelle (*Pipistrellus kuhlii*)], reproductive success in studied urban colonies is either lower than, or does not differ from, that in non-urban colonies (see Chap. 5). It is worth noting, however, that offspring may be weaned earlier in urban areas, which could improve juvenile survival and population recruitment [3–5]. At the population level, urbanisation can lead to reduced population genetic diversity that then puts populations at risk of extirpation because they are less resilient to change ([5], see Chap. 2). For bats, the consequences of living in urban areas are not straightforward and may not be as beneficial as purported. Instead, urban areas may be ecological traps (e.g. see Chaps. 5 and 6) where bats that *appear* equipped to live there may suffer fitness costs and even extirpation. To establish if urban areas are ecological traps, studies should determine if bats are attracted to these areas despite reduced fitness.

2 How Do Bats Compare to Other Mammals in Their Response to Urbanisation?

Bats are among the most common mammals in urban environments [5], possibly because flight allows them to move among isolated habitat patches more easily than less mobile mammals like terrestrial carnivores or arboreal species [6–9]. For example, fruit bats [e.g. Egyptian fruit bats (*Rousettus aegyptiacus*)] target stands of fruit trees in urban areas that are often more densely concentrated, more diverse, and more productive compared to stands of non-urban fruit trees [10]. Mobility could also allow species that are typically sensitive to habitat loss and degradation, such as tricolored bats (*Perimyotis subflavus*) and various species of mouse-eared bats (*Myotis* spp.), to obtain some of the resources they need in urban areas while seeking other resources beyond a city's limits [9, 11–15] (see Chap. 8). Additionally, flight may allow bats to navigate around urban stressors like traffic and artificial lights, just as subterranean mammals use burrows to escape disturbance [5, 6, 8].

While flight may allow bats to access habitat patches in urban environments, it may also pose trade-offs in adapting to urban living. Flight limits how much weight bats can carry and is energetically demanding, which constrains how much energy bats can devote to offspring production and care [16, 17]. Bats therefore produce few offspring (typically one annually) that develop slowly (3–4 months) but are long-lived [7, 16, 17], whereas most small urban-dwelling mammals produce relatively large litters with shorter lifespans [5]. Because they have longer generation times, species like bats with slow life histories are often slower to adapt to changes in their environment [5]. This might explain why, despite their incredible diversity of more than 1400 species, only 6.5% of bat species have been documented in urban areas [5]. By contrast, less diverse taxa like hyraxes (Hyracoidea), carnivores (Carnivora), and insectivores (Eulipotyphla) with faster life histories have relatively higher representation in urban environments with 20%, 10.6%, and 9.5% urban species in each order, respectively [5].

Species that persist in urban environments typically possess traits like the ability to produce larger litter sizes, flexibility in weaning age, and behavioural flexibility that predispose them to respond to urban stressors and possibly offset higher mortality rates in urban environments [5]. The small percentage of bat species studied in urban areas provides equivocal support for this argument (e.g. see Chap. 3). Just as primates found in urban environments typically produce more offspring (twins) than non-urban primates, bat species commonly found in urban environments also produce larger litters than non-urban bats [e.g. two pups (noctule bats, *Nyctalus* spp., and pipistrelle bats, *Pipistrellus* spp.) and up to four pups (hairy-tailed bats, *Lasiurus* spp.) [5]]. That said, some commonly cited "synanthropic" bat species such as little brown myotis (*Myotis lucifugus* [18]), big brown bats (*Eptesicus fuscus* [19]), and greater short-nosed fruit bats [*Cynopterus sphinx* (see Chap. 5)] may have lower reproductive success (but note the trends are not always straightforward) when compared to populations of non-urban conspecifics. It is difficult to determine age at weaning for bats because volant juveniles may continue to nurse, but timing of parturition and age at volancy could be suitable proxies. Parturition is earlier in urban *P. kuhlii* compared to non-urban conspecifics, which the authors suggest may improve recruitment by increasing juvenile survival and allowing females to breed in their first year [3]. Timing of weaning/fledging was not recorded in this study; thus, it is not clear if earlier parturition also means earlier fledging. In some years, weaning occurs later resulting in a longer lactation period in urban *M. lucifugus* compared to conspecific non-urban populations [18], which is consistent with urban-dwelling carnivores and rodents and may facilitate brain development and learning [5].

Mammals, including bats, may overcome some of the challenges of urban living through behavioural flexibility that allows them to tolerate risk and disturbance better than their non-urban equivalents [5, 9]. Generalist bat species, like molossid species and *E. fuscus*, have adjusted to habitat loss by living in artificial structures and foraging in a variety of patches of green space [[15] (see Chaps. 3, 6, and 8)]. Just as diet diversity has allowed a variety of carnivores (Carnivora) and ungulates (Cetartiodactyla) to persist in urban environments [5], urban *R. aegyptiacus* have broader diets that include the diverse fruits typically planted in urban areas and not found in the diets of rural conspecifics [10]. By contrast, specialists [e.g. *P. subflavus* and northern myotis (*M. septentrionalis*)] only occur in urban areas where they can access patches of high-quality habitat [1, 11–15, 20]. Santini et al. speculated that behavioural plasticity is related to relatively larger brains typical of many urban-dwelling mammals, including bats [5].

3   Impacts of Urban Environments on Hibernation and Migration

Like many mammals, bats in seasonal, semi-tropical, or temperate habitats face periods of reduced food availability and increased thermoregulatory costs due to changes in ambient conditions. In response, some urban-dwelling mammals (e.g. rodents, hedgehogs, and some carnivores) build up fat stores and cache food to remain active during these periods, while others, including bats, rely on accumulated fat stores for hibernation and migration [21–23]. Arousal from hibernation is energetically expensive and may be triggered by human disturbance [21], so bats hibernating in urban areas (e.g. *E. fuscus*, in buildings) may be more subject to disturbance than those hibernating in rural areas. Bats can minimise these risks by migrating short distances to hibernacula outside urban areas or longer distances to winter roosts [24]. Occupation of urban areas may therefore be seasonal for many bat species.

Migration and hibernation phenologies are being disrupted by climate change and urban heat islands (UHIs), which are also interacting to affect local weather patterns (e.g. temperature and precipitation) so that urban winters are now warmer compared to adjacent rural areas (e.g. [25]). As a result, urban populations of some highly mobile animals that normally migrate, including several birds [e.g. American goldfinch (*Spinus tristis*), European starling (*Sturnus vulgaris*), evening grosbeak (*Coccothraustes vespertinus*), and purple finch (*Haemorhous purpureus*)], now live in urban areas year-round [26]. Recent reports from Europe suggest the same may be true for bats (e.g. *Nyctalus* spp.) that normally migrate long distances but are now detected year-round in some urban areas [27]. Whether the same is true of other urban bats remains to be tested. The fitness consequences of these shifts have also yet to be examined but warrant careful examination.

4   How Can We Support Urban Bats?

There are measures we can take to reduce the challenges faced by urban-dwelling bats, although some of the available tools have not yet been rigorously evaluated. Incorporating more green spaces into urban areas, including forested parks, tree-lined roads, and blue spaces, would buffer urban noise, offset UHIs, create commuting corridors, and provide possible roosting and foraging habitat (see Chaps. 7 and 8). Bat boxes can potentially replace lost natural and artificial roosts and enhance existing habitat (see Chap. 6). The use of diverse green space configurations and bat box designs should be investigated to accommodate foraging and roosting needs for a variety of species (see Chaps. 6, 7, and 8). In some countries, federal and municipal efforts are in place to improve urban canopy cover by limiting loss of existing trees and planting new ones [28], which could benefit bats. However, most targets involve planting young trees, which offer fewer benefits to bats that typically roost in, and forage among, older, mature trees. Citizen engagement is also key to supporting urban bat populations, but the most effective pathways to engagement depend on cultural context (see Chap. 10). Although most negative perceptions of bats are unfounded, there is legitimate concern around disease transmission (see Chap. 11). Educating the public about how to limit risks and the value of urban-dwelling bats is therefore key to supporting bat conservation in urban areas (see Chaps. 10, 11, and 12). Fortunately, there is some demand for more natural green spaces in cities so that meeting these demands may coincidentally support the needs of urban-dwelling bats. For example, a survey of 25,000 park staff from 27 Canadian cities revealed that 70% of cities are seeing higher demands for more natural green spaces to support biodiversity [28]. There are also an increasing number of conservation and stewardship efforts in Australia, Europe, and North America targeting urban bats specifically, including providing replacement habitat for bats upon eviction, public education, and other community-wide efforts such as citizen science or volunteering opportunities (see Chaps. 6, 10, 11, and 12). The question remains if these measures will translate to fitness benefits (i.e. survival and reproductive success) for bats and how this might vary across species. By contrast, bats are often considered a nuisance and receive little conservation support in much of the southern hemisphere, which may account for the relative dearth of work investigating urban-dwelling bats in these areas (see Chap. 10; [29]).

5   Future Directions for Research and Defining Urban Environments

Given the varied responses of bats to urbanisation, more studies are needed in regions underrepresented in the ecological literature, like much of the southern hemisphere. Future studies should present long-term, comparative data across species, as well as between urban and non-urban populations, and specific demographic groups within a population – including more detailed accounts of reproductive phenology. We also need measures of individual and lifetime reproduction, along with estimates of genetic diversity, to determine if urban areas are stable environments for bats or act as ecological traps. More studies of bat responses across an urban gradient (along with better defining "urban" – see below) would help predict bat responses to urban sprawl. A meta-analysis of five studies found bat activity was much lower in "intermediate" urban developments [1], where parasite loads also appear to be higher (see Chap. 4). We also need specific data about movements and patterns of habitat use across a large sample of individual bats, both urban and non-urban.

Captures in mist nets or harp traps, as well as traditional radio-tracking and acoustic monitoring, typically offer a snapshot of habitat use by bats but do not provide the fine details we need about what aspects of habitats are vital to bats. However, when used together, they can provide surprising details about habitat needs, as seen for the 7-g *M. septentrionalis* found roost-switching among mature trees in a 348-ha urban woodlot, as well as foraging in the woodlot and in a marsh <1 km away [20]. Automated telemetry across a network of receiving towers, such as the Motus Wildlife Tracking System, offers great potential for more detailed accounts of habitat use, including the 28 bat species currently being tracked (for more, visit Motus). GPS tracking systems, such as ATLAS and Icarus, provide nearly real-time tracking of animals and have provided detailed insight into the movements of *R. aegyptiacus*, revealing their use of cognitive maps (i.e. mental representations of space) to locate food and roost resources in an urban area and adjacent rural areas [30, 31]. Precise details of habitat use may provide a clear indication of how different urban stressors influence bats and how this could vary among species.

Where possible, explicit definitions of "urban" would be helpful as they are often absent from studies or inconsistent among studies, creating ambiguity and limiting potential for meaningful comparisons. Urban environments are often defined by "extreme" habitat modification created by impervious or built-up land, resulting in habitat loss, degradation, and fragmentation, as well as reduced biodiversity [1, 15]. How we define "urban" influences how we identify the boundaries between urban and non-urban environments, as well as estimates of urban population numbers (for a more detailed discussion, visit Urbanization - Our World in Data). Certainly, this is less concerning for qualitative comparisons across studies or for comparisons of populations or species within a single study. However, clearer definitions are needed if we are to compare responses across urban gradients and across taxa. Clear definitions are also needed when making predictions about how wildlife will respond to urbanising areas and to inform best practices for management. Similar considerations should be given to defining "green" or "natural" spaces in studies where these labels are used to denote a particular habitat type for comparison of bat diversity, activity, or fitness.

Literature Cited

1. Jung K, Threlfall CG (2018) Trait-dependent tolerance of bats to urbanization: a global meta-analysis. Proc R Soc B Biol Sci 285:20181222. https://doi.org/10.1098/rspb.2018.1222
2. Taylor M, Tuttle M (2019) Bats: an illustrated guide to all species, Illustrated edn. Smithsonian Books, Washington, DC
3. Ancillotto L, Tomassini A, Russo D (2015) The fancy city life: Kuhl's pipistrelle, *Pipistrellus kuhlii*, benefits from urbanisation. Wildl Res 42:598. https://doi.org/10.1071/WR15003
4. Lausen CL, Barclay RMR (2006) Benefits of living in a building: big brown bats (*Eptesicus fuscus*) in rocks versus buildings. J Mammal 87:362–370. https://doi.org/10.1644/05-MAMM-A-127R1.1
5. Santini L, Gonzalez-Suarez M, Russo D et al (2019) One strategy does not fit all: determinants of urban adaptation in mammals. Ecol Lett 22:365–376. https://doi.org/10.1111/ele.13199
6. McKinney ML (2002) Urbanization, biodiversity, and conservation: the impacts of urbanization on native species are poorly studied, but educating a highly urbanized human population about these impacts can greatly improve species conservation in all ecosystems. Bioscience 52:883–890. https://doi.org/10.1641/0006-3568(2002)052[0883:UBAC]2.0.CO;2
7. van der Ree R, McCarthy MA (2005) Inferring persistence of indigenous mammals in response to urbanisation. Anim Conserv 8:309–319. https://doi.org/10.1017/S1367943005002258
8. Munguía M, Trejo I, González-Salazar C, Pérez-Maqueo O (2016) Human impact gradient on mammalian biodiversity. Glob Ecol Conserv 6:79–92. https://doi.org/10.1016/j.gecco.2016.01.004
9. Ritzel K, Gallo T (2020) Behavior change in urban mammals: a systematic review. Front Ecol Evol 8:393. https://doi.org/10.3389/fevo.2020.576665
10. Egert-Berg K (2021) Fruit bats adjust their foraging strategies to urban environments to diversify their diet. BMC Biol 19(123):11. https://doi.org/10.1186/s12915-021-01060-x
11. Gehrt SD, Chelsvig JE (2003) Bat activity in an urban landscape: patterns at the landscape and microhabitat scale. Ecol Appl 13:12
12. Gehrt SD, Chelsvig JE (2004) Species-specific patterns of bat activity in an urban landscape. Ecol Appl 14:625–635. https://doi.org/10.1890/03-5013
13. Dixon MD (2012) Relationship between land cover and insectivorous bat activity in an urban landscape. Urban Ecosyst 15:683–695. https://doi.org/10.1007/s11252-011-0219-y
14. Coleman JL, Barclay RMR (2012) Urbanization and the abundance and diversity of Prairie bats. Urban Ecosyst 15:87–102. https://doi.org/10.1007/s11252-011-0181-8
15. Moretto L, Francis CM (2017) What factors limit bat abundance and diversity in temperate, North American urban environments? J Urban Ecol 3. https://doi.org/10.1093/jue/jux016
16. Chrichton EG, Krutzsch PH (2000) Reproductive biology of bats. Academic Press, New York
17. Barclay RMR, Harder LD (2003) Life histories of bats: life in the slow lane. In: Kunz TH, Fenton MB (eds) Bat ecology. University of Chicago Press, Chicago, pp 209–253
18. Coleman JL, Barclay RMR (2011) Influence of urbanization on demography of little brown bats (*Myotis lucifugus*) in the Prairies of North America. PLoS One 6:e20483. https://doi.org/10.1371/journal.pone.0020483
19. Patriquin KJ, Guy C, Hinds J, Ratcliffe JM (2019) Male and female bats differ in their use of a large urban park. J Urban Ecol 5:juz015. https://doi.org/10.1093/jue/juz015
20. Thorne T, Matczak E, Donnelly M et al (2021) Occurrence of a forest-dwelling bat, northern myotis (*Myotis septentrionalis*), within Canada's largest conurbation. J Urban Ecol 7. https://doi.org/10.1093/jue/juab029
21. Speakman JR, Thomas DW (2003) Physiological ecology and energetics of bats. In: Kunz TH, Fenton MB (eds) Bat ecology. Chicago University Press, Chicago, pp 430–490
22. Geiser F (2013) Hibernation. Curr Biol 23:R188–R193. https://doi.org/10.1016/j.cub.2013.01.062
23. Lyman CD (1961) Hibernation in mammals. Circulation XXIV:434–445. https://doi.org/10.1161/01.CIR.24.2.434

24. Fleming TH, Eby P (2003) Ecology of bat migration. In: Kunz TH, Fenton MB (eds) Bat ecology. Chicago University Press, Chicago, pp 156–208
25. Prokoph A, Patterson RT (2004) Application of wavelet and regression analysis in assessing temporal and geographic climate variability: eastern Ontario, Canada as a case study. Atmosphere-Ocean 42:201–212. https://doi.org/10.3137/ao.420304
26. Bonnet-Lebrun A-S, Manica A, Rodrigues ASL (2020) Effects of urbanization on bird migration. Biol Conserv 244:108423. https://doi.org/10.1016/j.biocon.2020.108423
27. Kohyt J, Pierzchała E, Pereswiet-Soltan A, Piksa K (2021) Seasonal activity of urban bats populations in temperate climate zone—a case study from southern Poland. Animals 11:1474. https://doi.org/10.3390/ani11051474
28. McGrath D, Plummer R, Bowen A (2021) Cultivating our urban forest future: a value-chain perspective. FACETS 6:2084–2109. https://doi.org/10.1139/facets-2021-0076
29. Straka TM, Coleman J, Macdonald EA, Kingston T (2021) Human dimensions of bat conservation – 10 recommendations to improve and diversify studies of human-bat interactions. Biol Conserv 262:109304. https://doi.org/10.1016/j.biocon.2021.109304
30. Toledo S, Shohami D, Schiffner I et al (2020) Cognitive map–based navigation in wild bats revealed by a new high-throughput tracking system. Science. https://doi.org/10.1126/science.aax6904
31. Harten L, Katz A, Goldshtein A et al (2020) The ontogeny of a mammalian cognitive map in the real world. Science. https://doi.org/10.1126/science.aay3354

Index

A

Aeroconservation, 96
Aeroecology, 96
Anthropogenic urban infrastructure, 96
Artificial light at night (ALAN), 5, 6, 96, 99, 100, 103, 125, 127, 160
Artificial roosts, 10, 76–78, 86, 88, 89, 110, 144, 185
Attitudes, 140–143, 148–150, 158

B

Bat-borne viruses, 155
Bat box, 76–89, 110, 144–146, 185
Bat house, 76
Behavioural plasticity, 34, 184
Bioacoustics, 125, 130
Biodiversity, 20, 23, 26, 46, 47, 53, 57, 62, 68, 69, 99, 100, 124, 125, 127, 133, 140, 168, 169, 175, 176, 185, 186
Blue space, 108–117, 185
Bottleneck, 26, 27, 156

C

Chiroptera, 21, 157, 169
Cities, 4, 5, 7–10, 12, 14, 15, 23, 34–36, 38, 44–46, 48–50, 52, 53, 56–58, 64, 76, 83, 97–102, 108–117, 124, 125, 127, 128, 130–132, 140–150, 155–161, 163, 169, 170, 173, 175, 176, 182, 183, 185

Cognitive Hierarchy Theory, 141
Community-driven conservation, 87
Conservation action, 69
Cynopterus brachyotis, 23–28, 170, 174

D

Drinking, 4, 7, 13–15, 55, 101, 109, 112, 116

E

Ecosystem services, 143, 148, 158, 168–178
Ectoparasites, 44, 46–53, 57
Endoparasites, 44, 46, 53–57

F

Fast-flying bats, 34, 39
Foraging, 4–15, 23, 33–40, 44, 46, 54, 56, 57, 69, 82, 89, 97–100, 102, 103, 108–117, 133, 160, 169, 170, 173, 176, 182, 184–186

G

Genomic diversity, 23, 24
Green space, 9, 23, 101, 108–116, 124–126, 132, 140, 144, 145, 163, 172, 182, 184, 185
Green urban remnants (GR), 126

H
Habitat, 4–10, 12, 14, 20–22, 26–28, 33, 35,
         44, 50, 53–58, 62–69, 76–89, 96–103,
         108–111, 113–117, 124, 125, 127, 130,
         131, 140, 145, 154, 156, 159, 168, 169,
         172, 173, 175, 176, 182–186
Human-wildlife conflict, 154–164

I
India, 64
Insectivory, 11, 13, 168–172

L
Lesser short-nosed fruit bat, 23, 25, 27

M
Molossidae, 9, 34, 126, 128

N
Noise and light pollution, 4, 49, 102, 182

O
Overheating, 76, 83–86, 88, 182

P
Pollination, 28, 147, 148, 168, 169,
         172–176, 178
Preadaptation, 4, 9, 13, 15, 34, 37, 39

R
Reproductive success, 8, 63, 64, 68, 69, 76,
         78, 79, 81, 182, 183, 185

Research directions, 117, 169
Roost energetics, 10, 12

S
Seed dispersal, 148, 168, 169, 172–176
Social system, 37, 38, 40, 62, 63, 68
Spillover, 154–159, 162–164
Stress physiology, 4–15
Synanthropes, 45, 154, 157, 162

U
Urban airspace, 96, 98, 99
Urban exploiters, 34
Urban fauna, 132
Urban gradient, 46, 50, 58, 173, 186
Urbanisation, 4–7, 20–23, 26–28, 33, 34, 40,
         44–57, 62–64, 68, 69, 124–133, 154,
         156, 168, 173, 175, 176, 182–185
Urbanites, 140–146, 148–150, 156–161, 170,
         175, 176
Urban landscapes, 37, 39, 63, 69, 82, 89, 113,
         114, 116, 117, 131, 132, 168
Urban mammals, 168
Urban syndrome, 182

V
Vectors, 47, 50, 54, 172, 176

W
Water balance, 4, 10, 13–15
Wildlife disease, 154, 155, 158

Z
Zoonoses, 46, 47, 155, 164

Printed in the United States
by Baker & Taylor Publisher Services